Energy Storage

Energy Storage

Edited by
Yves Brunet

First published 2011 in Great Britain and the United States by ISTE Ltd and John Wiley & Sons, Inc. Adapted and updated from *Problématiques du stockage d'énergie* published 2009 in France by Hermes Science/Lavoisier © LAVOISIER 2009

ISTE Ltd
27-37 St George's Road
London SW19 4EU
UK

www.iste.co.uk

John Wiley & Sons, Inc.
111 River Street
Hoboken, NJ 07030
USA

www.wiley.com

Library of Congress Cataloging-in-Publication Data

Energy storage / edited by Yves Brunet.
p. cm.
Includes bibliographical references and index.
ISBN 978-1-84821-183-4
1. Energy storage. 2. Electric power supplies to apparatus. I. Brunet, Yves.
TK2980.E54 2010
621.31'26--dc22

2010022199

British Library Cataloguing-in-Publication Data
A CIP record for this book is available from the British Library
ISBN 978-1-84821-183-4

Printed and bound in Great Britain by CPI Antony Rowe, Chippenham and Eastbourne.

Table of contents

Foreword

Sources of energy: density of stored energy

Energy sources are all stored, whether on a geological scale or greater (the sun), and the stores are used up according to need (the idea of "renewables" therefore is only meaningful when considering human timescales). We can distinguish the primary source of fossil fuels that exist "naturally" and for which we only pay the cost of extraction, from secondary sources, which are man-made, and for which we must pay for both storage and extraction.

Sources	Unit of time
Biomass	Years
Oceanic thermal gradients	Hundreds of years
Fossil fuels	Millions of years
Tides/waves	Hours
Geothermal	Days - years
Thermal mass	Hours
Batteries	Minutes
SMES	Seconds
Capacities	Seconds
Hyraulic pumping	Hours

Table 1. *Time required to replenish sources (source: W.A. Hermann, Quantifying Global Energy Resources, Science direct, Elsevier 2005)*

Foreword written by Yves BRUNET.
We may also refer to the chapter "Energy storage: applications related to the electricity vector" by the same author, in *Low Emission Power Generation Technologies and Energy Management*, ISTE / John Wiley, 2009.

In this work we will primarily be interested in these secondary forms of storage.

The energy that can be exploited is not only stored in nature under various forms, but is also stored with very different densities (Figure 1).

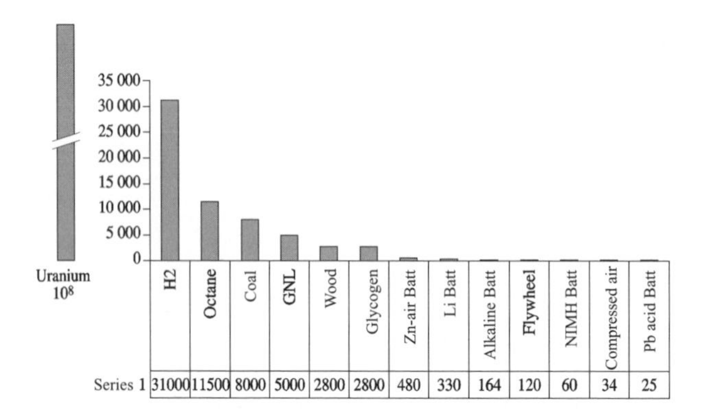

Figure 1. *The density of energy stored in materials or storage components varies greatly. The figure above shows the great advantage of fossil fuel sources over secondary storage sources. Nuclear sources are even more concentrated as we can obtain 10^8 Wh/kg from fission of natural uranium*

The range in the amounts of energy usage is such that it is good to consider a few simple applications: what can be done with 1 kWh?

We can:

– drive 1 km with a car that consumes 8 liters per 100 km;

– run a refrigerator for a day;

– light a house for an evening;

– make 200 g of steel or 100 g of plastic.

On average, the total amount of energy consumed in France, for each inhabitant, comes to 40 MWh/year, which is 4.5 kWh/hour per person.

Conversion of stored energy

Stored energy is released, according to target applications, either in the form of power (W), or in the form of energy[1] (J or Wh), which is sustained power over a

[1] 3,600 J = 1Wh, 1 MWh = 0.0857 toe (ton of oil equivalent), 1 tep = 11.7 MWh.

certain amount of time. Storage sources, which combine a quantity of stored energy with power that is instantaneously available, are often useful.

The storage strategy may lead to a range of different solutions (Figure 2).

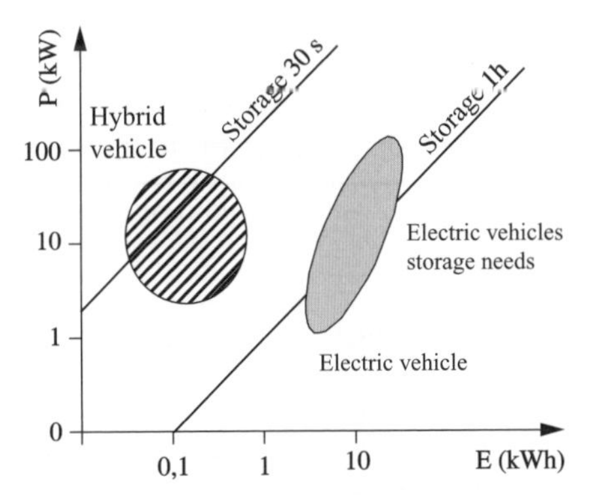

Figure 2. *The requirements of an electric vehicle. A hybrid vehicle essentially needs stored power, whereas an electric vehicle will need both power and energy. For a hybrid vehicle, the amounts required are of the order of 12 Wh/kg and 500 W/kg, with available energy of 300 Wh and available power of 10 kW over 2 seconds and a lifetime of 15 years*

Energy is brought to the user by an energy carrier, after transformation and conversion to the most suitable form possible for the target application. Electricity is one of these forms, without doubt the most flexible form known to this day (Figure 3).

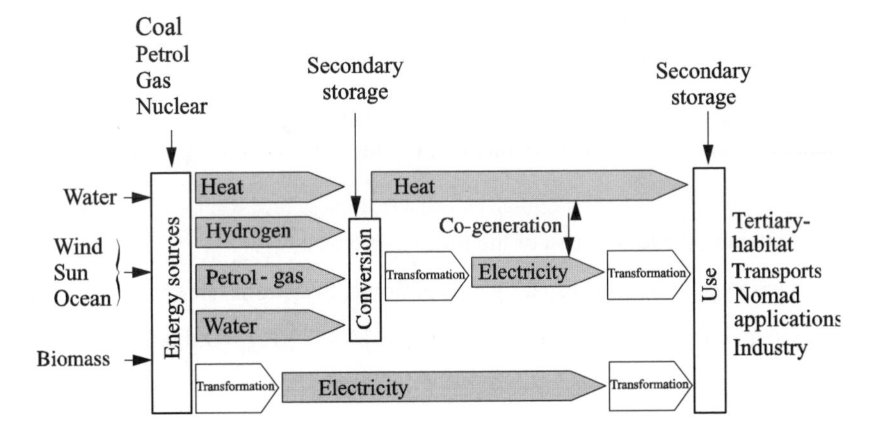

Figure 3. *Principal energy carriers*

The problem of energy storage is both technical and economic, and the solutions depend very much on the target applications (see Chapters 1-5). Regarding energy storage for technologies linked to the electricity carrier, this is not of immediate interest, particularly in the case of networks, and at least two opposing situations can be distinguished:

– onboard systems (mobile or portable applications, etc.), which carry their energy with them in order to ensure autonomous functioning, or pulsing systems for which storage acts as a "buffer" that releases the necessary high power;

– coupled systems (networks), which put into play high energy and high power.

The special case of pulsing systems[2]

A pulsing system stores energy and releases it in a very short time. In general, the energy is stored in electromagnetic form (an electric or magnetic field) and delivered in a very brief time (several milliseconds) as a result of a rapid switch. Therefore, for an amount of stored energy W, the power, P=W/t, can be very large.

In the case where energy is stored by a series of capacitors (Marx generators), several parameters are involved in the release of energy:

– the electrical characteristics of the storage circuit (R, L, C);

– the electrical characteristics of the charge impedance (R, L, C);

– the initial conditions;

– the characteristics of the switching system (R, L, t).

The voltage can reach several megavolts, for currents of several mega-amperes. Pulsing systems can be single shot or can go up to several kilohertz.

A capacitive system includes capacitances and a closing switch (V). An inductive system includes inductances and both closing (I) and opening (V) switches.

Switching devices can be of the following types:

– gaseous: pressurized spark gaps, ignitrons, thyratrons, etc.;

– semi-conductor: thyristor, GTO, IGBT, MOSFET, SRD diodes, etc.;

– solid: fuse.

---

2 I would like to thank Jean-Claude BRION (Europulse) for help with the editing of this section.

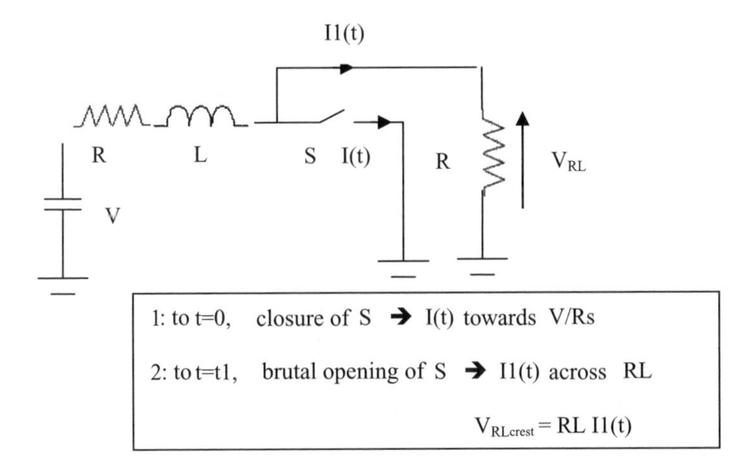

Figure 4. *Inductive storage for a voltage generator: the principle is to generate current in an inductance, and then to force the current to cross an impedance in a given instant. This technology requires a sharp opening switch*

In the case of capacitive storage, Marx generators enable a high voltage to be generated by charging capacitors in parallel and discharging them in series.

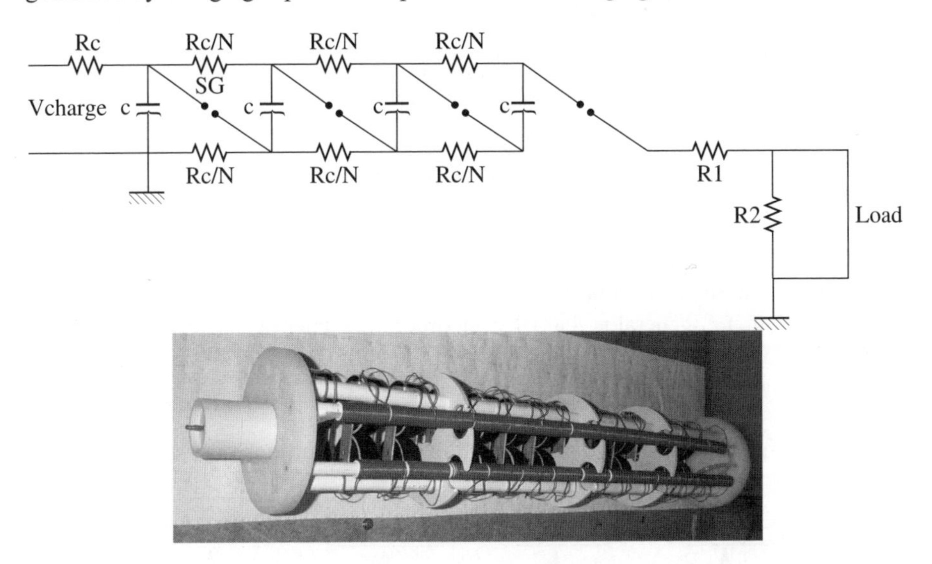

Figure 5. *Marx generator: diagram and photograph of a compact generator with 13 stages of 5.2 nF. 40 kV, 6 kJ/s. Vmax = 350 kV, mounting time = 15 ns, width of pulse = 50 ns, rate of repetition = 115 Hz*

Pulsed energy is used in several domains, industrial as well as research based. Among the applications using pulsed power-based systems, we can cite radar, particle accelerators, creation of very high magnetic fields, lasers, electric cannons (railgun) etc.

The special case of electrical networks

This case will be detailed in Chapter 1. Here, we outline the principal characteristics of storage in electrical networks. The storage problem takes on a greater level of seriousness when looking at electrical networks. As electricity is not readily stored in an efficient manner and in useful quantities, it is necessary to constantly adapt the power supplied to the power demanded, whilst recognizing that this fluctuates according to the time of day and season (Figure 6). Storage technologies break this link by allowing production and storage of electricity for later use.

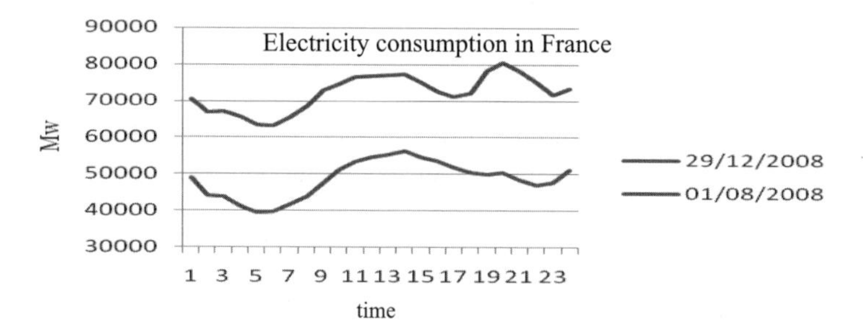

Figure 6. *Electricity consumption in France (source: RTE)*

It is therefore necessary to store energy in an intermediary physical form (mechanical, thermal, chemical, etc.) and to convert this stored energy into electricity (battery, generator, etc.) by incorporating energy converters, based on power electronics, whose efficiency (of the order of 80% to 90%) nevertheless has energy and financial costs.

In the energy chain, storage can be used in every one of these steps (Figure 7).

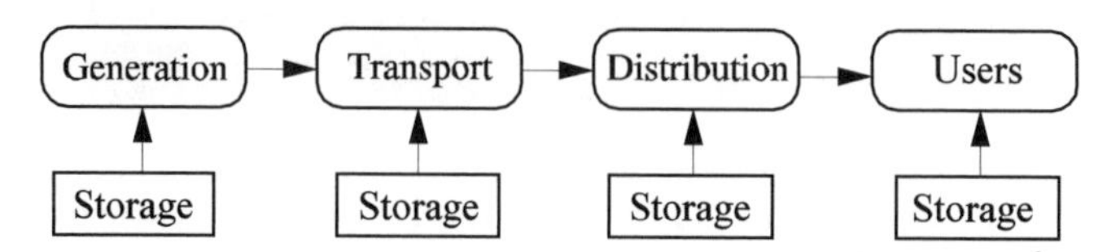

Figure 7. *Points where storage can be inserted within a network*

Storage technologies must demonstrate technical viability and economic interest. The cost of energy, linked to its variability according to time of day and of year (due to the supply and demand law in a market, which is increasingly open) and the difference between this cost at peak and off-peak times, are parameters that determine the degree of interest in adding storage. Storage is a means of adding flexibility competing with other factors:

– the value of the storage depends very much on the technology used and on its sizing compared to the predicted usage;

– the same type of storage can have a different value on different markets and for different agents;

– several factors have a strong influence on the value that agents can give to storage, such as the energy mix, the level of congestion on the network, etc.

A storage system can play different roles and can be, for example:

– a peak-time electric power station;

– a source of charge smoothing (harnessing transits over targeted work);

– a way to maintain the quality of the current, voltage, and frequency;

– a support to the network during downgraded function;

– a promotion permitting investment;

– a stabilizing function in a context where renewables have properly penetrated the market.

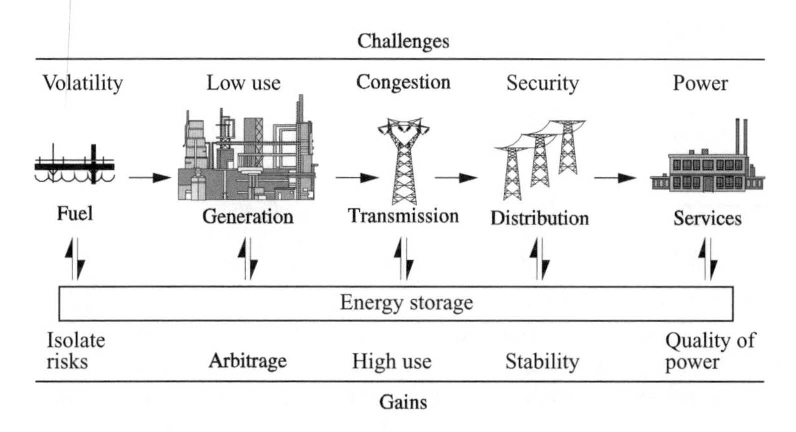

Figure 8. *Storage brings answers to problems in electrical networks (source Energy Storage, The Missing Link in the Electricity Value Chain: An ESC White Paper, Published by the Energy Storage Council, May, 2002)*

There are also intermediary situations (micro-networks, isolated systems, etc.), which often use intermittent energies (wind power, solar energy, etc.) for which the storage solutions must be studied according to technical and economic criteria. Storage, therefore, enables us to resolve the problem of intermittence of renewables by allowing us to:

– maximize the use of photovoltaic electricity;

– consume energy at the place of production and increase energy efficiency;

– increase the flexibility and efficiency of energy management;

– ensure safety of the user in the case of network outage.

Following the target applications, several technical and economic parameters (investment costs, energy or power densities, cyclability, impact on the environment, etc.) influence the choice of storage technologies (Figure 9). These different technologies will be detailed in Chapter 6 and later chapters.

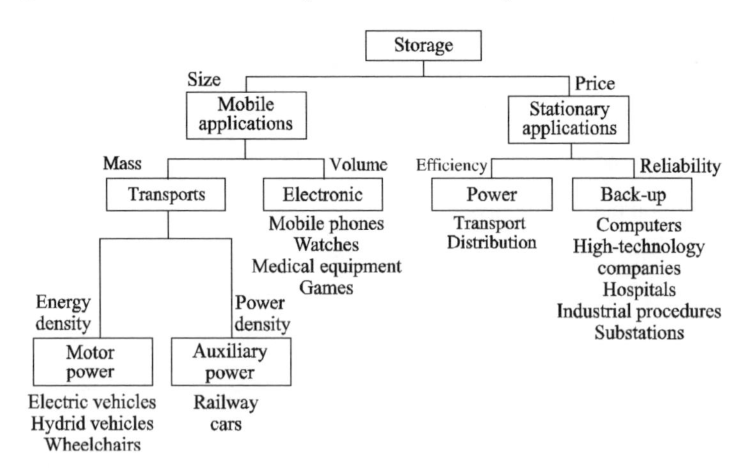

Figure 9. *Constraints and criteria for choosing storage technology based on applications*

If we look again at Figure 6, we see that using storage to account for increases beyond the average daily consumption of electricity leads to a requirement to store several tens of gigawatt hours. At the user level, the problems are different as the quantities of power are much lower, and it might be interesting to consider storage solutions closer to where they are needed[3] (Figure 10). Storage is a way to

3 EPRI 2.4 kW, 15 kWh Salt River Residential Photovoltaic-Battery Energy Storage System Project 1997.

guarantee the quality of the energy at user level (UPS, Uninterruptible Power System).

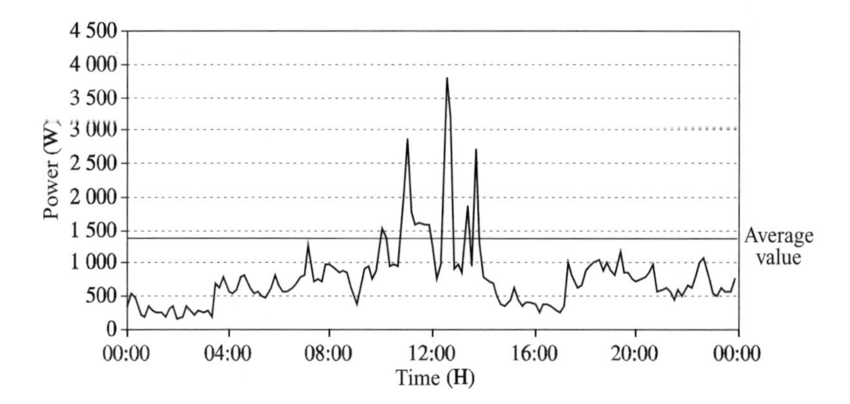

Figure 10. *Daily consumption for a house (November 2005[4]). Instead of installing 6 kW of power, corresponding to a standard power that is already higher than the maximum necessary, we could install 1 kW of power connecting to the network (average value = 780 kW) and linked to storage system of 4 kW,4 kWh, which would be able to deal with peak times*

However, the financial cost resulting from instantaneous interruptions or from prolonged interruptions (blackouts) of the electrical network is very important as the network today touches all sectors of the economy (it is calculated to be several tens of billions of dollars per year in the USA)[5] and this cost must be compared to that of the storage systems that could reduce the risks of interruption.

Power and energy must be globally managed using network management systems that use ICT (Information and Communication Technology) at the network-operator level, even more in the presence of distributed production. In addition to their traditional function of control-command, these systems are also capable of managing the entire production, storage, and charge using virtual power stations.

Storage technologies

The two tables below summarize the different storage technologies alongside their domains of application.

4 Doc GIE IDEA (Tuan Tran Quoc).

5 Communication J ETO EESAT 2004.

Technology	Gravity hydraulics	Underground compressed air	Electrochemical batteries	Circulation batteries	Heat and turbine

Energy density	1 kWh/m³ for a drop of 360 m	12 kWh pr m³ of underground space at 100 bars	Lead battery: 33 kWh/t Li-ion battery: 100 kWh/t	33 kWh/m³	200 kWh/m³
Achievable capacity	1000 - 100000 MWh	100 - 10000 MWh	0.1 - 40 MWh	10 - qq 100 MWh	1000 - 100000 MWh
Achievable power	100 - 1000 MW	100 - 1000 MW	0.1 - 10 MW	1 - qq 10 MW	10 - 100 MW
Electricity Efficiency	65%-80%	50% (with the support of natural gas)	70% per month in rapid discharge	70%	60%
Installations existing	100 000 MWh 1000 MW	600 MWh 290 MW	40 MWh 10 MW	120 MWh 15 MW	–
Cost E/kWh E/kW	70 to 150 600 to 1500	50 to 80 400 to 1200	200 (Pb) to 2000 (Li) 300 (Pb) to 3000 (Li)	100 to 300 1000 to 2000	50 350 to 1000
Maturity	Very good	Several examples throughout the world	Several examples throughout the world	Working prototypes in development	Planning stage
Notes	Site with altitude difference and water reserves	Underground site	Heavy metals	Chemical products	Independent of geographical constraints

Table 2. *Storage technologies of high capacity (source CEA)*

Technology	Inductive superconductor	Supercapacitor	Electrochemical	Flywheel	Bottled compressed air	Reversible PAC hydrogen

Energy form	Magnetic	Electrostatic	Chemical	Mechanical	Compressed air	Fuel
Energy density (only accumulator, without attached equipment)	1 to 5 Wh/kg	**10 Wh/kg → 60Wh/kg**	20 to 120 Wh/kg	1 to 5 Wh/kg	**8 Wh/kg (200 bars)**	300 to 600 Wh/g (200 to 350 bars) without PAC
Achievable or achieved capacity	several kWh	several kWh	several Wh to several MWh	several kWh to several 10 kWh	several kWh to several 10 kWh	N/A
Time constant	several seconds to 1 mn	several seconds to several minutes	several 10 minutes (NiCd) to several 10 hours (Pb)	several minutes to 1 hour	1 hour to several days (little autodischarge)	1 hour to several days (little autodischarge)

Table 3. *Storage technologies of average and low capacity (source CEA)*

Chapter 1

Energy Storage for Electrical Systems

1.1. Introduction

This chapter addresses the potential applications for energy storage in electrical networks or, more specifically, in "electrical systems". The term "electrical network" tends to refer mostly to transmission and distribution networks, whereas the more general term of "electrical systems" encapsulates the entire electric power supply chain, comprising:

– electricity generation, not only by centralized power stations (whether they be nuclear, fossil fueled, hydraulic, etc.), but also by smaller decentralized generation units (cogeneration, diesel, etc.), or from renewable energy (RE) sources (wind, photovoltaic power, etc.);

– the transmission and distribution networks, with different levels of voltage (from 400 kV for very high voltage transmission networks up to 400 V on low voltage feeders);

– electricity consumption by different types of customers connected to these networks: industrial, commercial and tertiary sectors, residential customers, etc.

In electrical systems, the need to maintain the balance between production and consumption of electricity at each instant has made energy storage an issue for a long time. In fact storage systems have been present for a very long time, such as, for example, pumped hydro energy storage known in French as STEP (*Stations de transfert d'énergie par pompage*/hydraulic pumping stations). However, the

---

Chapter written by Régine BELHOMME, Jérôme DUVAL, Gauthier DELILLE, Gilles MALARANGE, Julien MARTIN and Andrei NEKRASSOV

economic conditions for the majority of energy storage systems, comprising high costs, economic constraints related to the access to the grids, insufficient financial returns, etc., have prevented the level of development that would have been expected in this area.

However, the current situation and future evolution scenarios for the electric sector bring new perspectives on energy storage, and reasons for modification of the economic conditions include [JAC 08]:

– the need to reduce carbon dioxide (CO_2) emissions;

– the development and integration of intermittent RE;

– acknowledgment of the fact that traditional energy sources are dwindling;

– the rise in prices of fossil fuels that should result;

– volatility of the markets;

– networks being operated closer and closer to their limits and the difficulties encountered in developing further network infrastructures;

– technological evolution;

– regulatory evolution.

As a result there is a revival of interest and a large number of research projects are underway on different aspects of energy storage relating to electrical systems. In this context, every electricity system participant (or type of participant) has his own needs, and these lead to different applications for storage.

In this chapter, therefore, we review the main functions of electrical systems and the possible applications for storage. More precisely, we will consider energy storage:

– for the producer (section 1.2) and for the special case of integration of RE (section 1.3);

– for transmission (section 1.4) and distribution (section 1.5) networks;

– for the energy supplier or the retailer (see section 1.6);

– for consumers of electricity (section 1.7);

– for the balance responsible parties (section 1.8).

We will end with a brief summary (section 1.9).

1.2. Energy storage for the producer

The activity of "production" (or generation) consists of exploiting power stations and selling, at every instant, the produced energy on wholesale markets (for example, on spot markets where it will be bought by suppliers) or directly on retail markets (i.e. to the final customers).

In France, this activity takes place in the deregulated sector and, therefore, by participants that are in competition. "Pure" producers sell their entire production on wholesale markets. Integrated producers, which are both producers and retailers (see section 1.6), use all or part of their production in order to satisfy the energy needs of the portfolio "customers" on whom they rely commercially.

Whatever the nature of the markets considered by a producer, the volume and the revenue generated by its production are subject to different hazards:

– the volume sold at every instant by a producer depends on the availability of its power stations, on the size of the demand to be met, and on the competitiveness of its production costs;

– the produced energy, sold on a spot market or exported via interconnections, is paid for at a price depending on all the events and fluctuations occurring in electrical systems.

Faced with these uncertainties, the key issue for the producer is to optimize and secure its production and the associated revenues.

1.2.1. *"High-power energy storage" to maximize revenues associated with production*

The remuneration associated with the sale of produced energy fluctuates, especially according to hourly, weekly or seasonal variations in demand. In order to maximize production revenues, it is important for a producer to be able to sell the maximum amount of energy at times when the remunerations are most profitable.

Storage management allows energy to be stored when electricity prices are low, so that that energy can be sold when the electricity prices are higher. Therefore, storage is a lever that can allow a producer to increase the revenues associated with its production.

To fully profit from opportunities to carry energy forward, the storage capacities should be sized so that cycles of several dozen hours of use are possible and they should be of high power (order of magnitude: several hundreds of megawatts). Such capacities will enable energy to be carried forward from night to day, over the weekend, and on working days. Typically, the mature technologies that are targeted for this purpose are bulk storage systems, such as hydraulic pumping stations (STEP) and compressed-air energy storage (CAES).

Seasonal energy transfers from summer to winter can also present opportunities for increased revenue for a producer with hydraulic dams sized for storing major quantities of energy (cycles of several hundred hours of use). However, note that the sites (valleys), which are most suitable for developing such storage systems, are for the most part already used up in France.

In summary, by profiting from opportunities to carry forward production according to patterns in consumption (summer-winter, weekend-week, day-night), and by smoothing the residual load curve, energy storage allows a producer to maximize production revenues and to profit from new margins for exploitation. These new areas for exploitation correspond essentially:

– to a reduction in variable fuel charges: substitution of the most expensive fuels (fossil fuels), which are consumed at peak times by the least expensive fuels, which are consumed at off-peak times.

– to a better optimization of sale on the markets: selling more energy when the market conditions are more favorable;

– to the relaxation of the dynamic constraints affecting the operation and management of the generation park: smoothing the load curve enables better optimization of the generation unit commitment by limiting the weight of these constraints (for example, limiting the number of costly stop-starts of certain power generators);

– to a reduction in the level of CO_2 emissions which can be valued at the price of emission licenses: in particular in the case of inserting storage as a supplement to a generation mix with low CO_2 emissions in base-load (for example, hydraulic and nuclear) but with high CO_2 emissions at peak times (due to fossil fuelled stations).

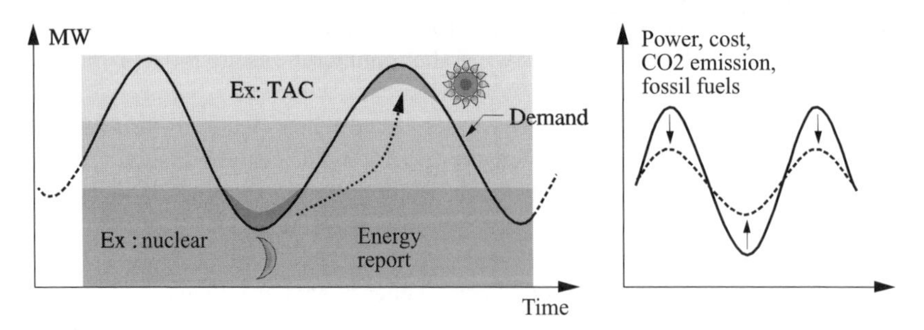

Figure 1.1. *Example of energy being carried forward in the daily cycle, as a result of a storage system based on a combustion turbine and the resulting advantages*

1.2.2. *"High-power energy storage" to alleviate physical and financial risks of production*

Availability of stored energy can help to limit the effect of unfavorable physical hazards, such as lack of available capacity in conventional production, cold fronts, lack of wind, etc. Energy storage constitutes an effective solution for alleviating these physical risks and enables producers to avoid having to take recourse in other forms of cover, such as buying energy on forward markets or investing in supplementary capacity.

In particular, storage enables to smooth the extreme peaks in the residual load curve, which reduces the need to invest in extra production capacity just to accommodate those load peaks that occur only a few hours each year. Therefore, storage permits the delay of, indeed even the reduction in need for, investing in new production capacity, which constitutes a saving for the producer.

In addition, availability of stored energy can cover financial risks, by limiting the level of exposure to the volatility of market prices. By providing the required energy to cover these risks, storage limits recourse to other ways of covering this risk, and this is yet another source of value of storage.

1.2.3. *Storage for ancillary services*

Depending on their size and their technical characteristics, storage systems, which are (or would be) at the disposal of a producer, can enable the producer's obligations to be fulfilled; for example, relating to regulation of frequency or network restoration. A succinct description of these services is given below.

1.2.3.1. *Frequency regulation*

The characteristics of frequency regulation depend very much on the country. As a result, there is much ambiguity in the terminology. For convenience, only the French case is considered in this chapter. We distinguish three types of frequency regulation: primary, secondary, and tertiary.

1) *Primary frequency regulation*: the goal of primary frequency regulation is to maintain real-time generation-consumption balance by acting directly and automatically on the governing systems of the generating units participating in the regulation. In particular, it is about stabilizing the frequency on a time scale of several seconds in the case of variations in frequency following an incident on the interconnected network.

Primary regulation of frequency is ensured due to the existence of an active power reserve (primary reserve) on a part of the generating units connected to the network, which therefore operate at reduced power. Therefore, the steady state power/frequency characteristic of the generating units contributing to the regulation is illustrated in the following figure.

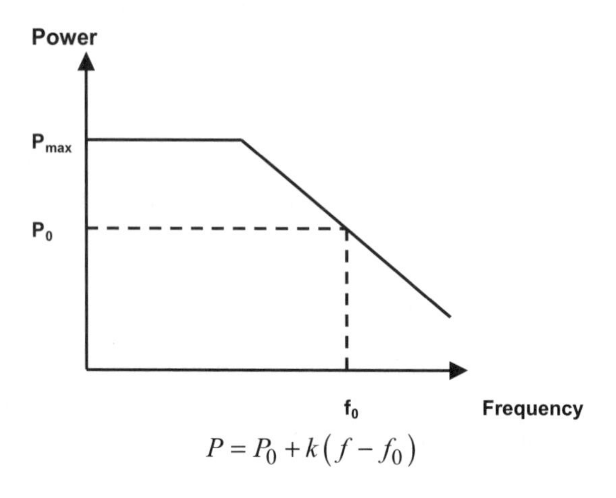

$$P = P_0 + k\left(f - f_0\right)$$

Figure 1.2. *Steady state power/frequency characteristic of a generating unit contributing to primary frequency regulation*

The primary reserve, established at the scale of the UCTE (Union for the Coordination of Transmission of Electricity) zone, must permit compensation for the reference incident, namely the sudden loss of 3,000 MW of production. In addition, in order to reduce the frequency gap in steady state, it is necessary to fulfill a minimum value for the ratio, $\frac{\Delta P}{\Delta f}$, where ΔP is the change in power of the generating units participating in the regulation and Δf is the change in frequency in steady state. This ratio is the regulating energy of the system, and it must be at least 18,000 MW/Hz for the first synchronous zone of the UCTE (Western Europe) and at least 3,000 MW/Hz for the second synchronous zone (Balkan countries). As a result, it is convenient to allocate the primary reserve to a sufficient number of generating units in order to guarantee the level of regulating energy required for the system.

In order to stabilize the frequency, the entire required reserve should be able to be supplied in 15 seconds for a gap in the production of less than or equal to 1,500 MW, and in 30 seconds for a gap of 3,000 MW (reference incident).

2) *Secondary frequency regulation*: this is an automatic centralized regulation that adjust the production of the generation units participating in the regulation in order to bring the frequency and the exchanges of power with neighboring electric networks to their target values. In contrast to the primary frequency regulation, secondary regulation is not a local regulation and requires sending of a signal to the generating units. This signal is calculated at the control center of the transmission system operator (TSO). Secondary frequency regulation is ensured thanks to the existence of an active power reserve (secondary reserve) on the generating units participating in this regulation. Only generating units of a sufficient size (greater than 120 MW in France) can participate in this regulation.

3) *Tertiary frequency regulation*: this is a manual regulation that serves to:

– feed the primary and secondary power reserves and to bring the frequency back to its required value when secondary regulation has not been able to do so (following a secondary reserve deficit);

– rebalance the system when there is slow growth in the gap between supply and demand.

Depending on the country, tertiary regulation can also be used to solve congestion on the transmission network.

Tertiary frequency regulation calls for tertiary active power reserves that can be mobilized at different time scales [RTE]. In France, these reserves are mobilized by a telephone call to the production facilities from the control center of the TSO.

The tertiary reserves are often associated with the balancing mechanism organized by the TSO. This is like a permanent call for tenders where balancing responsible parties (see section 1.8) submit "adjustment" offers, i.e. production offers upward or downward with some precise characteristics. The TSO then chooses the offers that fit the needs of the system in order to face the imbalances and ensure generation-consumption balance as well as system security.

1.2.3.2. *Restoration of the network*

After a total or partial collapse of the network, the objective of network restoration is to restore the supply as soon as possible, first for the priority customers, and then progressively for all customers and to return the electrical network to normal operation [RTE 04].

Restoration of the network comprises a number of steps and relies on generating units. Depending on their size and their technical characteristics, energy storage systems could contribute to restoration in the same way as generating units. In France, only generating units of more than 40 MW can participate in this service.

1.3. The special case of intermittent generation

1.3.1. *Contribution to frequency regulation in the absence of storage*

One of the problems introduced by intermittent generation (such as photovoltaics, wind power, etc.) is linked to its limited capacity to participate in ancillary services, and especially in frequency regulation. Due to its variable character and because of the rules for buying back this type of generation, it is customary to exploit renewable generation at the maximum available power, without participating in frequency regulation.

The lack of participation of wind power in frequency regulation has not been a problem until now, while the level of market penetration was weak. Nevertheless, the increase in installed power makes it necessary hereafter to provide ancillary services, as wind power tends to substitute the production from classical generating units which were used to ensure frequency regulation.

1.3.1.1. *Regulatory evolution*

Wind power is not controllable, but the creation of reserves for wind farms is more economically penalizing than for traditional means of production, for which the reduction in production can lead to a drop in the consumption of fuel.

Nevertheless, regulatory evolution is on the way, as the growing importance of wind power production could compromise the security of the electricity system. This is why, in countries where the penetration level of wind energy is particularly high, from now on "grid codes" foresee the participation of wind energy in frequency regulation, even if this participation is complicated by the intermittence of the resource.

This is the case, for example, in Denmark, where the grid code includes advanced functionalities where wind farms participate in maintaining the frequency, and in Ireland [ESB 04], where the grid is weakly connected and the level of penetration is especially significant, fluctuations in wind energy production have a great impact on the frequency of the grid.

1.3.1.2. *Regulation modes*

It is possible to implement functionalities of primary frequency regulation at the level of the wind farm controller. According to the characteristic given in [SOR 05], at nominal frequency the wind farm operates at reduced power compared to the power that is available. Therefore, the offshore wind farm at Horns Rev is equipped with a control system that allows its turbines to operate at reduced power and to contribute to primary frequency regulation. There is a boundary on either side of the nominal frequency within which requests for regulation are prohibited during normal operation, and the active power follows a linear characteristic when there is a significant gap in frequency.

The order for a reduction in power during normal operation is generated by the control system of the wind farm, using the maximum power available at the level of each turbine as the basis. This order can be generated in different ways (constant order for reduced power, maintaining a constant gap between supplied and producible power, limiting the power gradients of the wind farm, etc.).

1.3.1.3. *Limitations*

Insofar as the primary energy source is not controllable, any voluntary reduction in power leads to suboptimal functioning from the point of view of the producer.

In addition, the intermittent character of wind energy production does not allow a guarantee of the possibility for increasing power, as the maximum power of each turbine is directly linked to the availability of wind energy (note, however, that the differences in wind power production in the different parts of the electricity network allows smoothing the variations in wind power production at a time scale corresponding to the one of the primary frequency regulation).

This fact leads to investigation of the possibility of using storage systems associated with wind power production for primary frequency regulation, in order to:

– economically optimize the operation of the wind farm; the turbines would be led to operate at a power that is closer to their maximum power while the power at the grid connection point would be modulated by the storage system as a function of the frequency of the network (at the level of a fraction of the nominal power of the turbine, and over short periods);

– better guarantee the availability of the reserve for power increase, thanks to stored energy.

1.3.2. *Contribution of storage to power/frequency regulation*

1.3.2.1. *Impact of aggregation*

A study for the US Department of Energy [KIR 04] evaluates the supplementary power required to contribute to frequency regulation for a wind farm. The farm considered in the study is large (138 turbines producing 103 MW in total), and so it is possible to see the impact of aggregation on the participation in primary frequency regulation.

To do this, the farm was split up into four groups of turbines and the power required for the regulating service was evaluated for each group. The results are given in Table 1.1. We see that the 4.8 MW of stand-alone regulation capacity required by the wind plant is about 65% of the 7.5 MW of regulation that would be required if the four sections were to address regulation independently. The author of the study deduced that aggregation has a positive effect on regulation.

	Groups of turbines				
	A	B	C	D	Together
Number of turbines	30	39	14	55	138
Power in MW	23	29	10	41	103
Power required for frequency regulation (MW)	1.8	2.2	1.0	2.5	4.8

Table 1.1. *Impact of aggregation at the scale of a wind farm, from [KIR 04]*

On the one hand, the impact of wind energy on frequency regulation on the scale of one large control zone is hardly noticeable. Therefore, the primary reserve of other means of production will not be affected by fluctuations in wind energy production. On the other hand, the positive effect of aggregation tends to reduce the need for storage on the scale of a large wind park (i.e. a group of wind farms). Therefore, the contribution of wind power plant to primary frequency regulation should be estimated at the scale of the electrical network, and not at the scale of a single wind farm. However, this poses the question as to what type of electricity system participants is going to install the storage systems (localized storage at the level of a farm, or centralized storage managed by a participant that has to be specified, e.g. such as a production aggregator).

In addition, in the conclusions of the study, the author indicated that the desired characteristics for the storage correspond well in this case with the performance of inertial flywheel systems (good cycling capacity, short response time, and reduced discharge time). In this way, it appears that the required storage is of the type "storage in power", when the available power is the most important parameter.

1.3.2.2. *Strategy for using storage*

Traditional generating units contributing to primary frequency regulation should have a power/frequency characteristic that conforms to the characteristic reported in Figure 1.3. The power is modulated around a reference value, and if the frequency remains constant and equal to 50 Hz, the production should not change.

It is evident that this cannot be verified on the scale of a single wind farm with a very fluctuating production. In order to ensure an equivalent contribution with a wind farm, the associated storage system should fulfill two functions:

– modulation of power for frequency regulation;

– maintaining the reference value (for frequency this is 50 Hz) independently from variations in production.

These require a storage size that is probably unrealistic. In contrast, it is currently known that instantaneous wind power production can be considered to be constant on the scale of the electric network for durations in the order of a few minutes[1].

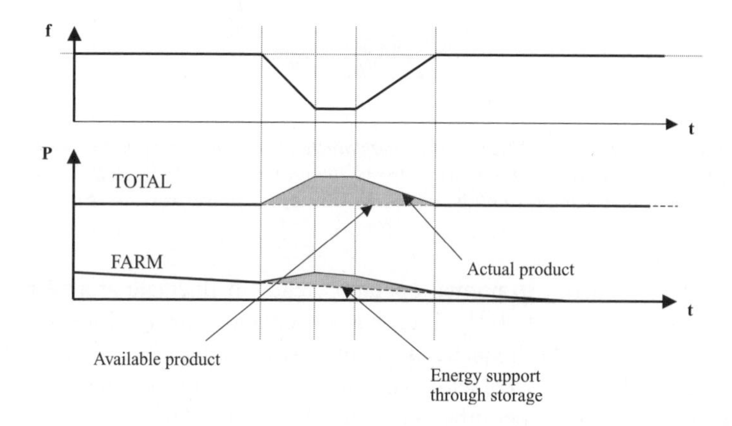

Figure 1.3. *Power modulation at the scale of a farm and a park*

In this way, the contribution to primary regulation of a wind park (i.e. a collection of wind farms) could be obtained by leaving the power of each turbine unchanged (at the maximum available power), but modulating the power at the grid connection point of the farm via the storage. The power for the entire wind park will then follow the P/f characteristic desired, thanks to the natural scattering between the different farms.

1.3.2.3. *Storage charge management*

Contrary to traditional generating units, primary frequency regulation is not limited by a variation range of power between some technical minimum and the specific power capacity of the limiter. The limit of the contribution of wind power to primary regulation will be linked to the charge state of the storage.

1 This explains why the impact of wind power on the volume of the primary reserve is usually considered negligible.

Every type of storage presents a minimal charge beyond which it is impossible to descend (risk of degrading the equipment). In addition, in order to use the same volume of reserve during upward and downward trends, it is convenient to maintain, as far as possible, a level of intermediary charge in order to avoid the critical situations reported in Figure 1.4.

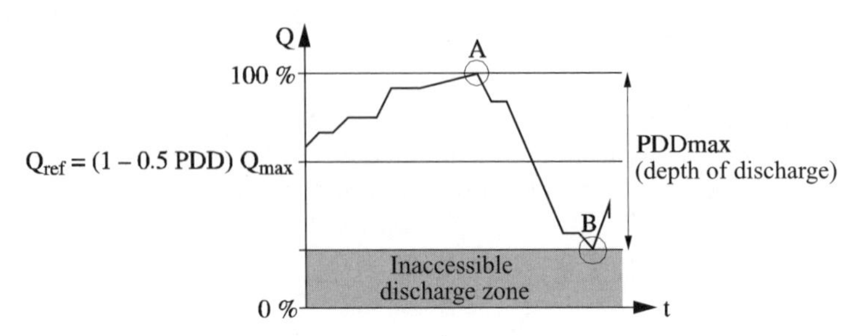

Figure 1.4. *Level of storage charge. At A, the charge is maximal; it is no longer possible to curtail production, whatever the power reserve available. Conversely, at B, it is no longer possible to increase production, no matter how much power reserve is available*

In fact, the frequency in normal operation tends to fluctuate around its normal value with a mean value of 50 Hz. Frequency fluctuations being of weak amplitude, and the mean value of the frequency gap being zero *a priori*, the state of charge of the storage should remain satisfactory and close to the reference value. Nevertheless, when there is significant perturbation of the frequency, the reconstitution of the storage charge may be compromised. It will then be necessary to determine to what extent the wind power's reserve will be able to be reconstituted both in power and in energy.

The primary regulation cannot be effective unless the available energy allows the maximal power to be supplied, which may be required for regulation over a duration of at least 15 minutes.

1.3.3. *Other possible ancillary services for power storage*

1.3.3.1. *Secondary and tertiary frequency regulation*

Up to this point, only primary regulation has been considered. Nevertheless, the contribution to secondary and tertiary regulation seems currently to be beyond the scope of wind power production.

On the one hand, the wind power itself has an impact on the reserve volumes required, and the intermittence of production can no longer be neglected. Therefore, the power of the storage devices used must cover not only the power required by the regulation, but also some of the variation of the wind power.

On the other hand, this reserve should be free to be used over longer durations and therefore, it needs higher energy. Therefore, secondary and tertiary regulations bring up applications, such as load deferral, that are inaccessible to power storage devices which would be dedicated to primary regulation.

1.3.3.2. *Voltage regulation*

Storage could favor voltage regulation via modulation of reactive power at the connection point. Admittedly, wind turbines more and more often present voltage regulation functionalities, but this practice would enable contribution to the voltage regulation even in absence of production. In addition, it would be necessary to determine how much such a system would permit the range of regulation of reactive power at the connection point to be extended. The possibilities for voltage regulation are detailed later.

1.3.3.3. *Other applications*

The use of a storage system of reduced capacity equally permits the anticipation of applications to do with power quality. Rapid fluctuations of production can induce transitory effects on the distribution voltage. Also, it would be pertinent to evaluate the capability of storage systems for smoothing these transitory effects, and to see to what extent such a function would interfere with primary regulation of frequency.

Another application could be the contribution to the restoration of the network after a blackout. One of the questions posed by the recovery from an incident is the capacity of the power generators to maintain voltage and frequency in the restored zone. Wind power generation occupies a growing place in the generating facilities, and so the regulation capacities could be a decisive point to regulate the voltage from the distribution side.

1.4. Energy storage for transmission systems

Depending on the mode of operation, storage systems can behave either as generating units or as loads (consumption). Therefore, in principle, they can ensure the same services on transmission networks. The majority of applications for transmission networks have already been considered in the two previous sections. These will not be detailed again here, and whenever needed we will refer to those sections.

1.4.1. *Control of investments and congestion management*

Depending on their size and their technical characteristics, energy storage systems could be used when charging or discharging, in order to control the flow of power on transmission lines. They could contribute in this way to maintaining the flows at values below the maximum acceptable.

This service can be used by the control center of the transmission system to solve congestion on the transmission network (as is already done using generating units) and to postpone certain investments. Moreover, the energy storage system can appear as a solution in cases where difficulties in the development of network infrastructures arise (for example, strong local opposition).

It takes cycles of a few hours duration to smooth over peaks of power in the transmission lines. If the duration of use of the storage system is much longer, reinforcement of the network will be inevitable. Regarding congestion management, depending on the country, the control center of the transmission system can call upon the tertiary reserve or upon the balancing mechanism (section 1.2.3.1).

1.4.2. *Frequency regulation and the balancing mechanism*

Just as for generating units, storage systems can participate in frequency regulation. This application has already been the subject of section 1.2.3.1 where participation in primary, secondary, and tertiary frequency regulation was presented.

We will not copy the description here. We recall only that participation in frequency regulation requires having access to reserve volumes upward as well as downward, i.e. the reserves should be used to inject power into the network as much as to store power from the network. Therefore, particular technical characteristics and specific operation modes might be required.

Moreover, storage systems can also participate in balancing mechanisms. These were also briefly described in section 1.2.3.1 and will be discussed again in section 1.8, which discusses energy storage for balance responsible parties who submit offers on the balancing mechanism (or on the balancing market – see section 1.8). Generally, a minimum volume is required in order to have access to the balancing mechanism. In France, the minimum volume of an offer is 10 MW [RTE 09].

1.4.3. *Voltage regulation and power quality*

Voltage regulation is not the primary function of a storage system. Other more efficient and less costly specialized systems exist for these kinds of applications.

Nevertheless, when a storage system is present on the transmission network (to fulfil another function), voltage regulation can be promoted using the alternator or the power electronics interface with which it may be equipped. However, resizing may be necessary.

Following the example of frequency, three types of voltage regulation exist: primary, secondary, and tertiary:

– *Primary voltage regulation* is an automatic local regulation that maintains the voltage at a given point of the network at a regulated value. In order to carry out this task, the generating units are equipped with automatic voltage "regulators". Other types of equipment on the transmission network can also carry out this function, for example static reactive power compensators, STATCOMs, etc. In France, every generating unit linked to the transmission network should be equipped with a primary voltage regulation system.

– *Secondary voltage regulation* is an automatic centralized regulation that coordinates the actions of the voltage regulators of the generating units that contribute to the secondary regulation in a such a way to control the voltage schemes of predefined zones. In France, only generating units linked to voltage levels from 225 kV to 400 kV are required to contribute to secondary voltage regulation.

– *Tertiary voltage regulation* is a manual regulation undertaken by network operators in order to coordinate the voltage scheme between different secondary regulation zones.

In the same way, even if it is not its primary function, a storage system can contribute to improving the power quality on a network, insofar as it has the technical capacity (for example, using its power electronics interface, if it is equipped with such a device).

1.4.4. *System security and network restoration*

Beyond their participation in regulating frequency and voltage, energy storage systems can also contribute to the security of the electricity system when they are in charging mode, and in particular:

– to load shedding: in case of "frequency collapse", when ordinary regulation actions do not enable control of the downward trend, the TSO cuts off the loads when the frequency reaches certain thresholds. In France, four power cut thresholds are fixed: 49 Hz, 48.5 Hz, 48 Hz, and 47.5 Hz. A curtailment level (the volume of load to be cut off) is associated with each threshold. Storage systems that are charging can be cut off, and this is exactly what happens with STEP operating on the network;

– to maintain voltage stability: in the case of voltage collapse, some action on the load is again a possibility that can be used by the TSO. The storage could contribute to this action.

Finally, as was already mentioned in section 1.2.3.2, after a total or partial blackout of the network, depending on their size and their technical characteristics, energy storage systems could contribute to the restoration of the network in the same way as generating units.

1.4.5. *Other possible applications*

Islanded networks: in some circumstances, parts of the transmission system can operate as islanded networks. For example, in the case of a blackout or a long-duration power cut following problems on the network, functioning as a separate (or islanded) network may be authorized while waiting for complete re-establishment of the network.

Islanded networks are generally less stable and are susceptible to frequency and voltage fluctuations that are more significant than for the normally operating interconnected network. However, the TSO should ensure the supply-demand balance in real-time; in this framework, because of the flexibility offered by the existing storage systems, they can contribute to maintaining this balance and the stability of the separate network.

1.5. Energy storage for distribution networks

Traditional use of storage in distribution networks consists of providing emergency power to certain infrastructures in the network. A good example is batteries in a substation for control/command systems and electrical protection. In the following sections we describe other more innovative services that may be facilitated by storage units installed through the distribution network.

1.5.1. *Storage advantages in planning phase*

1.5.1.1. *Storage for load smoothing*

In order to manage investments by the network operator, load smoothing is one of the services considered for storage and even, on a wider scale, for distributed energy resources (including distributed generation and load management).

When the nominal capacity of a piece of network equipment (or a fraction of this capacity considered as admissible) is likely to be exceeded in the near future due to the expected increase of the load the conventional solution consists either of reinforcing the existing infrastructures or in building new ones. Due to the standardization of network equipment, the corresponding increase in capacity is often much higher than what is needed in the short-term and therefore this leads to an "under-exploitation" (or a low usage factor) of the new assets for many years.

The use of a storage unit downstream from congested network components may be a flexible temporary solution. As shown on Figure 1.5, the charging of the storage unit is carried out during off-peak hours in order to constitute an active power reserve which will be injected into the network during peak hours and will contribute to reducing the maximum currents flowing in the upstream network components. This will avoid them becoming congested. Apparent transits may be reduced by acting on the active power and, more marginally, using a local reactive compensation according to predefined profiles or in a closed loop using a measure of instantaneous transits.

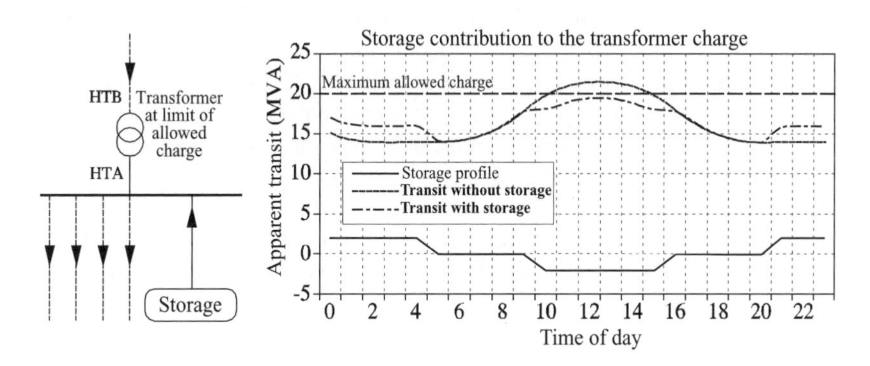

Figure 1.5. *An illustration of storage for load smoothing*

Overcoming consumption peaks transitorily in this way enables investments in capacity to be avoided or at least delayed. For example, [NOU 07] presents the setting up of a sodium sulfide (NaS) storage unit of 1 MW/7.2 h on the 12 kV network of American Electric Power in 2006. The installation enables peak shaving for a 46 kV/12 kV transformer of 20 MVA, which is close to surcharge, and in this way prolongs its use by several years. The creation of a new station, with heavy investment, can be deferred in this way. A reduction of thermal constraints on the network assets has also been observed, possibly leading to a longer life duration of these assets.

At the end of the investment deferral time, the necessary reinforcement is put in place. Storage is either maintained onsite, or moved to another point in the network. The peak shaving service can prove particularly useful in certain cases:

– when some kind of restriction (environmental, legal, local opposition) prevents or delays a project, at the risk of degrading the quality and/or the continuity of the power supply;

– in order to avoid reinforcements, which would be necessary for temporary connections, as for construction sites, for example. By extension, any installations that are only present on the network for up to 5 to 15 years (the typical lifetime of storage technologies) can be concerned.

The deferral of investment will require storage to be placed downstream from the constrained elements, which leaves some freedom of choice in the localization of the storage devices, so they can be located to allow shared use of services. Parameters to be taken into account are numerous, however, and should be studied on a case by case basis: availability of land, accessibility, communication demands, acceptability, possible sharing of resources, etc.

Placing the storage to deliver electricity to locations as close as possible to where it is needed (smoothing at the source of fluctuations) leads to the maximum number of new network assets. However, a more centralized storage (for example, situated near the HV/MV substation), benefits from the effect of aggregation, and is easier to manage than having many diffuse units: it allows the same level of deferral on upstream installations for a smaller sizing [MAR 98].

When technical thresholds are reached on the network, the decision of the DSO (distribution system operator) to implement a solution does not need any economic justification. However, if several solutions are technically possible, the choice is made with regards to an economic optimum. Given the different strategies that are possible (reinforcement, renewal of part or all of the system, doubling the feeder), it is usual to choose the option that minimizes the costs at a given time, taking into account the different expenses, such as investments, network losses, maintenance of the network assets, blackouts, etc. The economic interest of the peak shaving service should, therefore, be defined in this framework, comparing the "storage" option with other possibilities.

1.5.1.2. *Contribution of storage to voltage control*

Contrary to frequency, which is a global value on the network based on the instantaneous balance between production and consumption, voltage is an essentially local value that should be correctly regulated in order to ensure normal function of equipment. Therefore, the system operators should respect the regulatory and contractual constraints concerning the value of the voltage that they deliver to end users on their networks. Various technical methods are used by the utility to satisfy these needs; for example, the on-load tap changers of the HV/MV transformers and the off-load tap changers at MV/LV distribution substations.

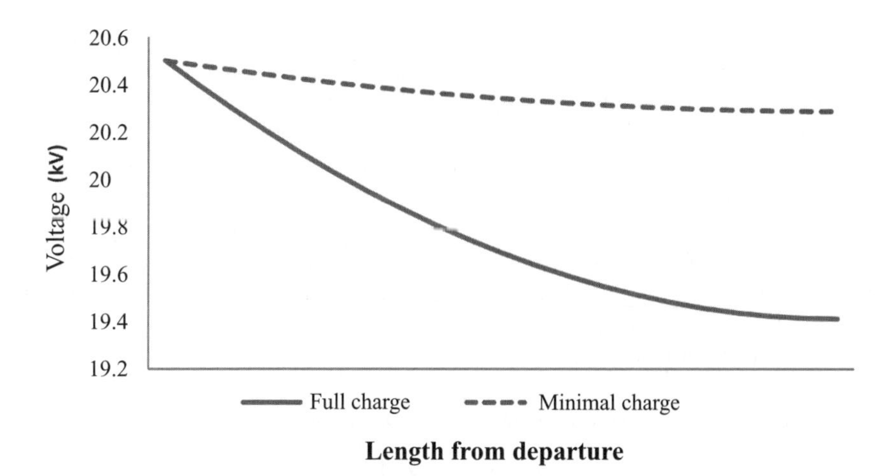

Figure 1.6. *Graph of voltage along a MV line*

Due to the impedance of the lines, power transits lead to voltage drops along the feeders, which can be quantified, for example, by using the following simplified expression:

$$\frac{\Delta U}{U} \approx \frac{RP + XQ}{U^2}$$

where U is the voltage, R and X are the resistance and reactance of the line, and P and Q are the active and reactive power transits.

In distribution lines, in the absence of decentralized production, the voltage progressively falls from the HV/MV substation to the extremities (Figure 1.6): the network planner specifically sees to avoid low voltage at peak times.

In recent years, many research studies have reported on possibilities for distributed generation (DG) to contribute to the regulation of voltage in order to increase the levels of penetration of DG or to reduce the economic cost of new connections. In an identical way, distributed storage can inject active or reactive power on an existing feeder in order to ensure a satisfactory level of power quality during peak times.

Thus, storage is an alternative to investments that would have to be undertaken in order to maintain the quality of the supply within contractual limits, such as, for example, the increase in the cross-section of conductors over a section of a line in order to reduce the impedance, and therefore, to reduce the voltage drop across it.

As an example, references [VRB 05] and [VRB 07] present a vanadium-based redox flow battery of 350 kVA/8 h (250 kW/250 kvar) used by PacifiCorp (United States) in 2003 to maintain the voltage in a 25 kV line that was exceptionally long. Following complaints concerning the quality of supply and faced with the impossibility of answering requests for new connections due to the constraints on too low voltage, the operator evaluated different solutions, such as new means of compensating for the reactive power, reinforcement of the existing line, or replacement of the line. In particularly restrictive environmental conditions (natural park), it was storage that was finally chosen. The system is installed at the midpoint of the line and ensures the level of voltage via peak shaving according to preprogrammed profiles.

1.5.1.3. *Storage to support a network in a degraded operation mode*

As a general rule, each MV line departing from a substation have one or more possibilities to be connected to another line from the same substation, or from another substation in order to rapidly restore the supply to end users in the event of a fault. The back-up scheme modifies the topology of the network and therefore affects the values of the electricity variables (current, voltage, powers): it should therefore be taken into account at the planning phase. The verifications for the MV level in "N-1 conditions[2]" are undertaken for peak consumption and focus on the voltage profile and the currents that circulate through the network assets.

From all the evidence, this method leads to the conservation of margins that can prove to be elevated compared to the effective needs of the normal situation, and this to be able to face events that are by their nature punctual and relatively rare. It is necessary to over-invest in a potentially important way: the study of alternative options, such as storage, can therefore be justified.

Distributed storage can bring support to situations involving degraded operation in a way that reduces the eventual constraints on equipment. It makes at least two types of contributions (potentially simultaneously): shaving the load peak seen by the rescuing substation and provision of local support of the voltage profile in such a way as to keep the consumers connected either to the rescuing substation or the rescued substation within admissible voltage limits. So, this service is close to the ones mentioned previously and it can even be considered as a particular case.

1.5.2. *Other possible applications*

In principle, the benefits linked to peak shaving and regulation of voltage only require a targeted operation on a fixed number of days of the year (at peak times),

2 Loss of 1 component.

with a profile that is predictable a few hours or a few days in advance. Supporting the network in the case of a degraded operation is a relatively exceptional case, and simply requires sufficient energy reserves to be permanently maintained in case there is an incident.

As a result, outside the time periods that these services are concerned with, it is possible to exploit a storage unit to maximize the benefits for the DSO. For example, the following complementary advantages may exist:

– any service that charges the storage device at off-peak hours in order to discharge it at peak hours presents the potential for *reducing network losses* linked to their quadratic nature. The potential gain, which depends on various parameters such as the form of the load profile or the impedance of the network, can be taken into account as a deduction of the losses in the storage unit itself, whose volume is greater in the large majority of cases;

– distributed storage, via power electronics, can contribute to *compensation of reactive power* performed at the heart of distribution networks. The value of such a service can be estimated through the avoidance of investments in capacitors banks at the substation. Besides, the power electronics interface can also help the operator of the network to respect the different commitments made to the end users regarding the *quality of the supply* (active filtration);

– finally, distributed storage could lead to *restoring voltage in a part* of the distribution network, following an incident. This service either involves timely employment of mobile storage units used like fuel generators as needs dictate, or addition of local voltage restoration to the services offered by a stationary storage unit in order to realize a complementary valuation. One such situation is in the case of a zone that has marked problems of continuity of supply for which conventional solutions (loops, reinforcement) are difficult to set up. The interest in using storage on site is even stronger if the zone concerned is difficult to access during hazardous weather. The restoration of the supply could be supported by the storage by itself or linked to other local resources.

All this being said, the underlying technical problems involved in taking over a pocket of consumption using storage are numerous and relatively complex. In particular they include power quality, security, and the reappraisal of the distributor's current practice in order to take into account situations where consumers are supplied by an islanded section of the network using distributed energy resources.

1.6. Energy storage for retailers

Commercialization (retail) involves the sale of energy and associated services, with offers adapted to the needs and expectations of private and business customers. As with production (section 1.2), this activity belongs to the deregulated sector in France: the electricity market has been open to all customers since July 1, 2007. In order to satisfy all the customers, the participants running this activity (suppliers or "retailers") must define a supply plan that is adapted to the provisional load curves of their customers.

The integrated participants, combining retail activities and production activities, can supply all or part of the energy sold to their customers using the production capacities they rely on. Pure "non-integrated" retailers can only use the markets to ensure their supply: bilateral contracts or OTC (over the counter) negotiations, buying standardized products on the forward market, buying/selling on the spot market.

Whatever the sourcing strategy being considered, the volumes and costs of sourcing are subject to different hazards:

– supply volumes can depend on meteorological or economic hazards. For example, the energy needs of residential customers who are equipped with electric heating depends on the temperature. The consumption of industrial customers depends on the hazards associated with their industrial processes (breakdowns, economic situation, etc.) or with markets (arbitrating customers on the spot market).

– the unit cost of sourcing can depend on conditions (price and available quantities) of access to the markets, especially for a pure retailer that does not have production capacity to rely on. An integrated retailer is equally subjected to hazards, notably production hazards (for example, availability of the power stations, volume of intermittent energy).

Faced with these uncertainties, it is important for the supplier/retailer to reduce and secure his sourcing costs.

1.6.1. *Energy storage to reduce the cost of sourcing*

Storage management enables deferrals to be organized between off-peak and load peak times. This means allowing sourcing to occur in periods where the demand is greatest and, therefore, most expensive to meet using less expensive energy that was stored when there was a lower demand.

In this way, storage constitutes a lever that allows a supplier to reduce his sourcing costs. In order to fully play this role and profit as much as possible from the opportunities for deferral that are presented, the storage capacities required must be sized in such a way as to lead to cycles of several tens of hours of use, and they must be of high power (order of magnitude: several hundreds of megawatts). Mass storage technologies such as STEP and CAES are an answer to this issue.

Note that some specific uses such as electric water heaters can be shifted, within certain limits. Therefore, commercialization and offer management (for example, price incentives), which influence the investment in these uses, constitute a lever, similar to storage, enabling efficient organization of energy shifting, reducing the sourcing cost for a supplier.

1.6.2. *Storage to secure the cost of sourcing*

The availability of stored energy constitutes a reserve that reduces the amount of exposure to the unfavorable conditions of sourcing: high prices and lack of liquidity on the markets, high marginal costs of production or limited security margins.

In this sense, storage is a tool that permits management of the risks of "price and volume" that weigh on the sourcing of a retailer.

1.7. Energy storage for consumers

The services discussed in the next section mainly concern industrial or commercial end users. To a lesser extent, some applications can (or could in the near future) interest residential consumers even if the additional complexity, the storage system dimensions and the possible risks perceived are inhibiting factors for them.

1.7.1. *Storage for peak shaving*

Among the services for consumers, the literature often underlines peak shaving (see in particular [MAR 98], [EYE 04] and [NOR 07]), the value of which comes from the principle of electrical energy price setting. The price billed to the customer is the result of:

– one part that is proportional to the subscribed power ("the subscription");

– one part that is proportional to the consumed energy.

However, the subscribed power is a "maximum value", which in practice is only attained very rarely. From the customer viewpoint, this is about reserving a service and under-using it or, otherwise, overpaying with respect to energy needs.

The principle of peak shaving consists of leveling the load profile of a customer in order to reduce the subscribed power. In order to do this, storage recharging is done when the load of the customer is weak and storage discharge is synchronized with calls for high power, as illustrated for a fictitious case in Figure 1.7.

The interest in realizing such an operation evidently depends very strongly on the form of the load profile of the customer and on the different contractual elements such as the cost of subscribed power and the modes of invoicing for eventual overruns. According to the references [MAR 98] and [OUD 06], the most favorable cases correspond to short peaks whose occurrence is predictable in advance, in such a way as to limit the capacity of the storage to be installed (and therefore the cost of the leveling system).

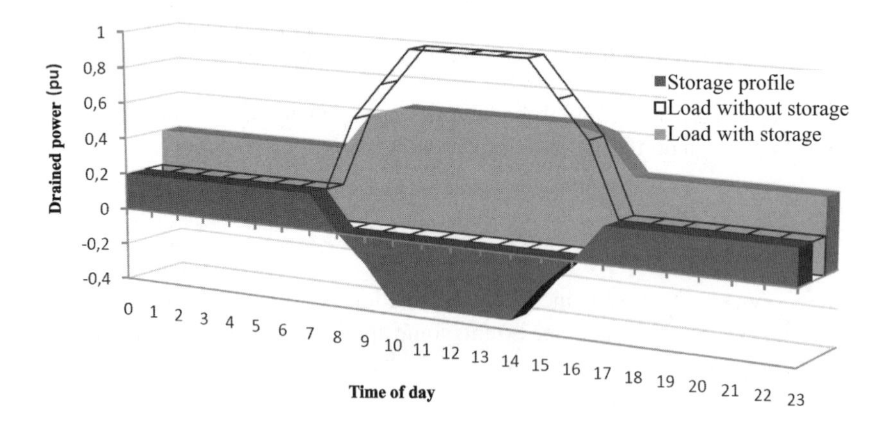

Figure 1.7. *Peak shaving with the help of storage*

The value of peak shaving is discussed in [EYE 04] and [NOR 07] in an American context. In addition, Oudalov *et al.* [OUD 06] studied the real case of a Swiss industrial customer for whom the subscription part represents 57% of the electricity cost, with a fixed premium of €89/kW. The optimal solution using storage would consist of a lead battery of 130 kW for 65 kWh – a system that would seem profitable after analysis (€32,000 of benefits realized over 20 years, with return on investment after 12 years). However, in practice, the peaks are not exactly known in advance and the real system would certainly have to be oversized as a result. This leads to a much less profitable balance sheet, even an unfavorable one if the storage unit is only required for the peak shaving. Another study of a very simple case is described in [MAR 98] and confirms that a leveling system could prove to be profitable under certain conditions, in an industrial context.

Mutualizations are possible in order to increase the value of the smoothing, such as gains in quality/continuity or else compensation of reactive power. Moreover, this service may naturally hold the benefits of a postponed consumption, if the peak takes place during the full-price hours. These different functions of the storage for customers are introduced later.

1.7.2. *Storage for deferral of consumption*

For the customer, this service consists of realizing some form of arbitration on the purchase price of his electricity (comprising supply and transportation) using the hourly differences in the tariffs. The storage charging is realized during off-peak hours (HC) at price C_1 and discharging is realized during peak hours (HP) at the higher price C_2 (Figure 1.8).

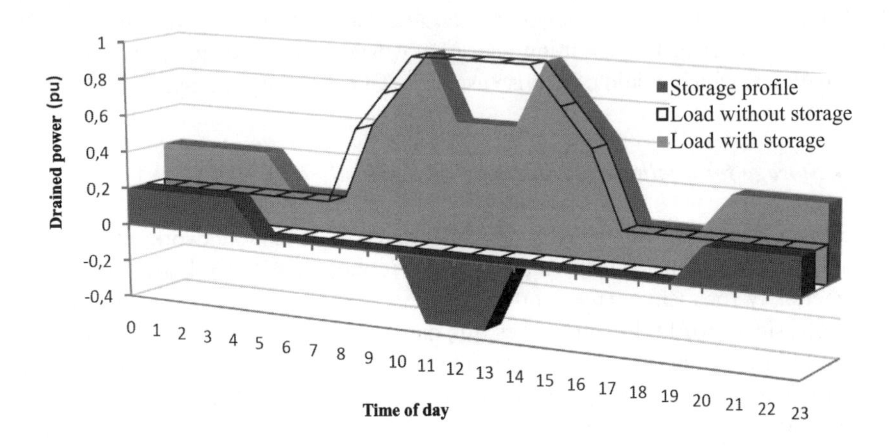

Figure 1.8. *Deferral of consumption via storage*

Schematically, taking into account the global efficiency η, of a storage unit, an opportunity only presents itself if the cost avoided by discharging storage compensates the additional expenses linked to buying stored energy. This means:

– cost of energy charged at off-peak hours: $\Delta C_{bought} = C_1 E_{charged}$

– cost avoided via discharging at peak hours: $\Delta C_{restitution} = C_2 \eta E_{charged}$

– opportunity is possible if $\Delta C_{restitution} > \Delta C_{bought}$ meaning $\frac{C_1}{C_2} < \eta$

Whatever happens, it is evident that the value of the load deferral service depends on the pricing scheme of the supplier. With a significant Peak/Off-peak ratio ($0.32/kWh versus $0.10/kWh), the calculations performed in [EYE 04] on real data lead to an estimated profit of $140/year for each kilowatt of storage installed (6 hours discharge time, η =0.8).

In the French case, the Peak/Off-peak ratio is less than the factor of 3 given in [EYE 04], and so the potential gain is relatively reduced. However, for a storage unit that would fulfill several functions, this service would have the advantage of at least partially compensating for the cost of losses in the storage unit. In addition, the foreseeable benefits could increase in the future with contracts that further exploit peak effects.

From the environmental viewpoint, the interest in deferral of consumption is in the transfer of lower CO_2-emitting energy at off-peak times to supply consumption at peak times, which would otherwise use more CO_2-emitting energy.

1.7.3. *Storage for quality and continuity of supply*

Storage can opportunistically be substituted onto the network at times of occasional disruption of the supply. This service has been known and used commercially for many years in order to reduce the consequences of power cuts in the professional world. In some cases, the damage is caused by the power cut itself, whatever its duration (in general electronic devices in general, possible loss of data). The solution consists of using very dynamic means that are able to replace the power supply for a cycle of the electric signals (a cycle of the sine wave). The power support brought by these means may be longer than a cycle or may be interrupted after having allowed for the controlled switching off of certain equipment and the completion of backup tasks. In other cases, it is principally the duration of the power cut that poses a problem (cold chain): it is important to have a backup electricity supply source that can start without response time restrictions and which has guaranteed autonomy.

In addition, the storage unit at the customer's site can be used to filter perturbations coming from the network, in such a way as to protect a process which requires a specific power quality. This may involve eliminating the voltage drops, for example, in order to protect a sensitive charge.

Uninterrupted power supplies (UPS) that are available commercially integrate all possible storage technologies. These systems exist for a very large range of applications: power up to 20 MVA, supply support time of a few seconds to several hours (with the possible support of a backup diesel generator), etc. Some models of active filters can equally be found on the market, but these systems are still quite new.

The amount of money that a customer could consent to invest in a service of this type is linked to the damage that could be suffered as a result of a poor power quality or power cuts. This amount is very dependent on the process and on the work material, etc. The accounting can prove to be even more difficult to do for residential end users and for the value they give to their consuming habits.

The literature mentions a fair number of tangible realizations using storage to ensure the quality and continuity of electricity supply, for example:

– the reference [ROB 05] describes a few commercial successes of storage for power applications (a few minutes at most) such as, for example, an uninterrupted power supply system of 15 MVA (12.5 MVA initially) for a semiconductor factory belonging to ST Microelectronics. During the 4 years after its set up in August 2000, this installation has taken over the supply for more than 100 occurrences of power supply disturbances, including power cuts of a duration of up to 20 seconds. For the company, the economic interest of a system of this size (directly connected to the MV substation of the factory) is in the replacement of a multitude of smaller low-voltage systems;

– at the other extreme of the spectrum of discharge times that are possible for storage technologies, Norris *et al.* [NOR 07] describe an experiment with two NAS modules (50 kW/7.2 h and 250 kW in a 30 second pulsing regime) tested in a commercial building of AEP in Ohio. Different applications which have a value for the consumer (notably, peak smoothing and consumption deferral) are combined in a service that ensures the temporary protection of loads against voltage drops, micro power cuts, and brief power cuts. The presented results are very conclusive: over a period of 4.5 months (from February to mid-June 2002), all the unacceptable phenomena were correctly handled by the installation (25 events in total, see [NOR 07] for the details and the criteria for classing the perturbations).

1.7.4. *Reactive power compensation*

In France, over the period November to March, consumers that are supplied on medium voltage level or in low voltage >36 kVA are invoiced for the reactive energy consumed beyond $\tan\varphi=0.4$ at the respective prices of c€1.77kVArh and c€1.86/kVArh (values given in the second version of the document specifying the Tariffs of Use of the Public Networks or TURP in French).

Distributed storage, thanks to the power electronics used for its connection to the grid, can compensate locally for the reactive power consumed by the loads. An electricity system participant equipped with storage could make the installation profitable by utilizing this aspect. The economic gain realized in this way can be established based on the cost of the reactive power or that of the banks of capacitors, which would permit the customer to conform to its commitments to the Distribution System Operator (DSO).

1.8. Energy storage for the balancing responsible party (BRP)

The role of a BRP exists in many electrical systems across the world. This electricity system participant contributes, across the *balancing mechanism*, to ensure the security of the electrical systems in the context of the market.

Any moral person, whether or not they have means of production at their disposal, or electricity purchase/sale contracts, can become a BRP[3], by signing a "BRP" contract with the Transmission System Operator (TSO) [RTE 09].

Each BRP is linked to a balance perimeter, which integrates physical entities for energy injection and extraction and energy exchange contracts on electricity markets[4]. The BRP commits itself to provide the TSO in advance with the necessary information to allow the latter to operate the network;[5] the BRP commits himself to financially compensate the TSO for the adjustment of the possible imbalances between the declared and the realized[6] programs at the level of his balance perimeter[7].

The time step (e.g. hourly) for the imbalance calculation and the modalities for their financial compensation vary from one country to the other, but most of the time they depend on the regulation at the national level. In Europe, the tariff for the imbalance settlement is most often defined on the basis of the offers for production

---

3 Responsable d'équilibre (RE) is the French terminology.

4 The physical entities and the electricity exchange contracts can belong to the BRP, as well as to other participants on the electricity system.

5 Mainly programmes of production, extraction and/or exchanges of blocks of energy, established for entities/participants belonging to the balance perimeter of the BRP.

6 The imbalances are calculated *a posteriori* by the TSO on the basis of a process which reconstructs the injection/extraction flows.

7 The BRP can benefit from the aggregation of the different (upward and downward) individual imbalances of the entities/participant linked to his perimeter.

increase (upward) or reduction (downward) submitted to the balancing mechanism, which are used to compensate for the imbalances.

For example, Figure 1.9 reflects the method for calculating tariffs for imbalance settlements as used by RTE, the French TSO.

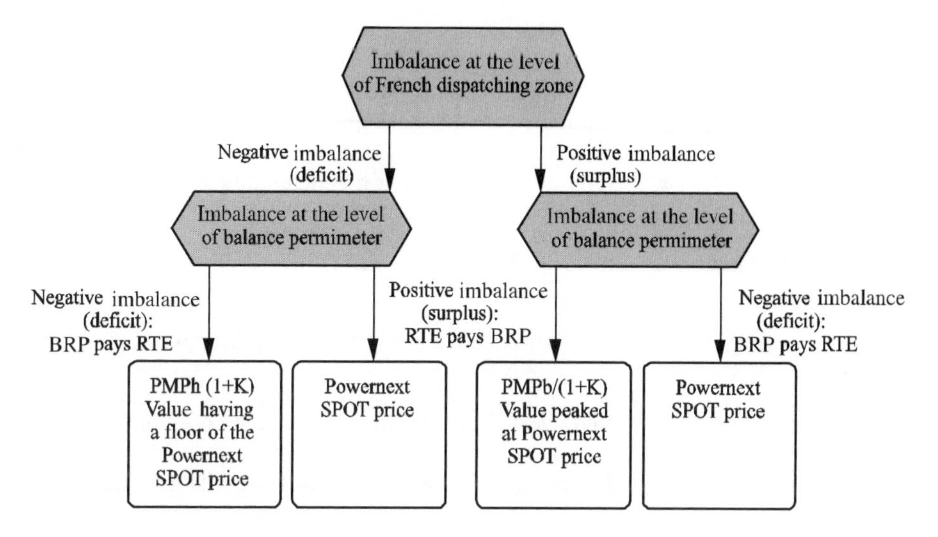

PMPb: average weighted cost of downward offers
PMPh: average weighted cost of upward offers
Factor K: coefficient that covers:

– the cost of the coexistence of upward and downward adjustments

– the remuneration of the availability guarantee on the balancing mechanism of fast-reserve offers

NB: for indication purposes only, in France the average weighted cost of downward offers (PMPb) is about 20% of the the SPOT price. Also, the average weighted cost of upward offers (PMPh) is about 120% of the SPOT price.

Figure 1.9. *Method for calculating tariffs for imbalance settlement used by RTE (France)*

We note that the tariff for imbalance settlement is sufficiently dissuasive so that the BRPs will be encouraged to minimize their imbalances. *This issue is important: to give an order of magnitude, the global cost of the German balancing mechanism in 2006 is estimated at 800 million Euros.* To satisfy this, the BRP can arbitrate between several actions at his disposal:

– improving the quality of forecast of injections/extractions in his balance perimeter (whether it is using his own means or those of its contractors);

– reinforcing the reliability of his production (if the lack of availability has a non-negligible effect on the imbalances);

– increasing the flexibility of his own production and of his own consumption or acquiring a supplementary flexibility from other suppliers.

Regarding this latter action, the BRP can use storage to reduce imbalances.

What will be the typical charge/discharge profile of a storage system for this purpose? Which storage technologies are the most appropriate?

Analysis of the profile of the imbalances for generation/load balance of the French electricity system in 2008[8] shows that:

– the maximum duration of an imbalance was: 70 hours downward, 32 hours upward;

– the average duration of an imbalance was: 6.4 hours downward, 5.3 hours upward;

– the maximum power used was 4,500 MW downward, 5,500 MW upward.

An example of the adjustment profile for the French system is given in Figure 1.10.

We note that the imbalances have a very pronounced *daily profile*. Therefore, the most appropriate storage would be the one for which the cycle of discharging the storage had an optimal duration of between 2 and 15 hours.

Without taking into account their maturity state, we can foresee the following technologies:

– hydraulic storage (pumped hydro station or STEP);

– compressed air energy storage (CAES, adiabatic CAES);

– electrochemical redox flow storage (vanadium-based electrolyte);

– hydropneumatic storage;

– thermal storage (at high or low temperature).

---

8 Details supplied by RTE, http://www.rte-france.com/htm/fr/vie/vie_mecanismehistorique.jsp.

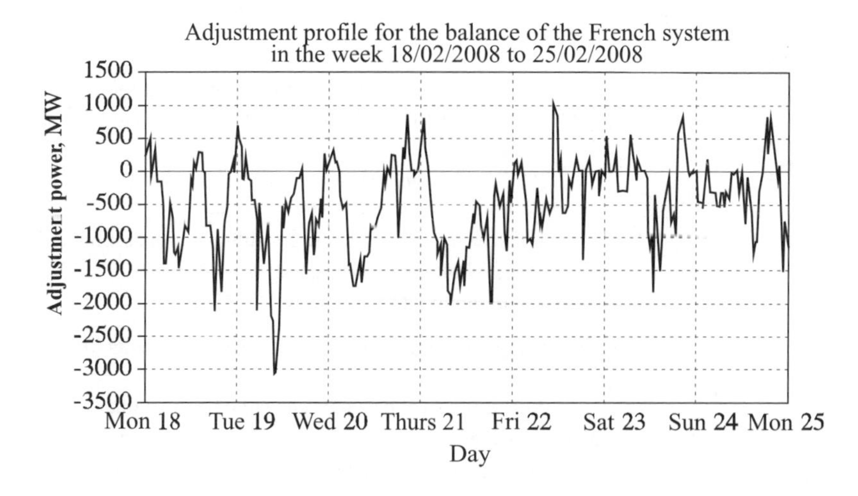

Figure 1.10. *Adjustment profile for the balance of operation*

In conclusion, the need to reduce imbalances can lead the BRP to call upon energy storage means that function with a daily cycle. The profitability of the use of storage to these ends in an electric system will depend on the cost of construction and use of the storage installations and on the level of the adjustment tariffs for the system in question.

1.9. Conclusion

In this chapter, we have reviewed the possible uses of energy storage for the main "participants" or for the main functions in electricity systems, i.e. for:

– the producer;

– the special case of integrating REs;

– the transmission and distribution networks;

– the energy supplier or retailer;

– the energy customers;

– and finally for BRPs.

Table 1.2 gives a summary. These uses will determine the sizing of the storage capacities and the technical performances required that will lead to a choice of storage technologies.

Participants or functions	**Possible uses of storage**
Producer	Maximizing revenues from the production as a result of deferral of production and smoothing of the residual load curve: – reduction of fuel costs; – optimization of sale on the markets; – relaxation of the dynamic constraints of operation at the level of the whole generation park; – reduction of CO_2 emissions. Covering physical risks and limiting the investments in production assets: – facing physical hazards (unavailability, weather phenomena, etc.); – shaving of extreme peaks. Covering financial risks: – limiting the degree of exposure to the volatility of market prices. Contributing to ancillary services: – regulation of frequency (primary, secondary, tertiary); – restoration of the networks.
Integration of renewable energy production	Contributing to: – regulation of frequency (primary, secondary, tertiary); – restoration of the networks; – voltage regulation; – maintenance or improvement of the power quality.
Transmission network	Managing network investments. Managing congestion. Frequency regulation and balancing mechanism. Regulation of voltage and power quality. Maintaining the security of the system. Restoration of the network. Maintaining the generation-load balance and the stability in islanded networks.
Distribution network	Peak shaving and managing network investments. Voltage regulation and compensation for reactive power. Support for the network in a degraded operation mode (backup scheme/configuration). Reduction of the losses on the network. Quality of supply. Restoring voltage to a pocket of the network after an incident.
Supplier/retailer	Reducing the cost of sourcing by profiting from opportunities for deferral. Securing the cost of sourcing by limiting the level of exposure to unfavorable conditions in terms of prices and volumes.

End users	Load peak shaving and reduction of the subscribed power. Deferral of consumption in order to profit from the best prices for electrical energy (for example, peak vs. off-peak hours). Quality and continuity of supply. Compensation for reactive power.
BRP	Reducing imbalances between the forecasts and the actual situation of the generation-consumption balance sheet for his balance perimeter: – compensating for errors in forecast of the consumption and distributed generation; – compensating for unexpected lack of availability of generating facilities; – increasing the flexibility of his customer portfolio (production and consumption).

Table 1.2. *Possible applications of storage in electricity systems*

Energy storage is currently at the center of many activities, both industrial and academic, across the world, from upstream studies (analysis of the storage needs of different participants, development of economic models for inserting storage in the energy system, etc.) to industrial projects (development of prototypes and commercial products) as well as experiments with existing storage technologies, etc.

However, even though energy storage technologies today have fulfilled many of the technical requirements, the economic profitability is not reached yet. This is why one of the fundamental questions regarding electrical energy storage concerns the estimation of its value. Indeed, taking into account the high cost of energy storage installations, the economic interest of storage has not yet been proven globally and is not conceded by all.

In particular, it is necessary to identify the conditions where the use of storage could today, or in a foreseeable future, justify itself economically: important evolutions in the cost of fossil fuels and/or explicit integration of the cost of emitting greenhouse gases, network congestion, special cases of isolated systems, etc. and, of course, evolution of the cost of energy storage itself as new technologies emerge.

Finally, the benefit value of energy storage by itself is not sufficient. It is important to take into account competitive solutions and their possible evolution, such as those in [BEL 08]: the management and control of load (or demand), centralized and decentralized production, network reinforcements, FACTS (Flexible AC Transmission Systems), market solutions, etc. For a given usage, the choice of the applied solution will result in a technical and economic comparison between the energy storage solution and the alternative possibilities.

1.10. Bibliography

[BEL 08] BELHOMME R., NAPPEZ C., NEKRASSOV A., "The flexibility challenge – creating new flexibility between consumption, generation and storage", *3rd International Conference on Integration of Renewable and Distributed Energy Resources*, Nice, France, December 10-12, 2008.

[CIM 05] CIMUCA G.-O., Système inertiel de stockage d'énergie associé à des générateurs éoliens, PhD thesis, ENSAM Lille, 2005.

[ESB 04] ESB National Grid, WFPS1 – wind farm power station grid code provisions, www.eirgrid.com/eirgridportal/uploads/Regulation%20and%20Pricing/WFPS1.pdf, July 2004, accessed October 2008.

[EYE 04] EYER J., IANNUCCI J., COREY G., Energy storage benefits and market analysis handbook. A study for the DOE Energy Storage Program, Sandia National Laboratories, SAND2004-6177, December 2004.

[JAC 08] JACQUEMELLE M., "Un vecteur de flexibilité pour le système électrique", *Conférence Le stockage d'énergie: quels enjeux pour le système électrique*, Le Printemps de la Recherche 2008, EDF R&D, Clamart, France, May 2008.

[KIR 04] KIRBY B.-J., *Frequency regulation basics and trends*, report of ORNL (Oak Ridge National Laboratory) for the DOE, http://www.ornl.gov/~webworks/cppr/y2001/rpt/122302.pdf, December 2004, accessed October 2008.

[MAR 98] MARQUET A., LEVILLAIN C., DAVRIU A., LAURENT S., JAUD P., "Stockage d'électricité dans les systèmes électriques", *Techniques de l'Ingénieur*, D4030, May 1998.

[MOR 06] MORREN J. *et al.*, "Inertial response of variable speed wind turbines", *Electric Power Systems Research*, vol. 76, pp. 980-987, 2006.

[NOR 07] NORRIS B., NEWMILLER J., PEEK G., NAS battery demonstration at American Electric Power. A study for the DOE Energy Storage Program, Sandia National Laboratories, SAND2006-6740, March 2007.

[NOU 07] NOURAI A., Installation of the first distributed energy storage system (DESS) at American Electric Power (AEP). A Study for the DOE Energy Storage Program, Sandia National Laboratories, SAND2007-3580, June 2007.

[OUD 06] OUDALOV A., CHARTOUNI D., OHLER C., LINHOFER G., "Value analysis of battery energy storage applications in power systems", *IEEE PES Power System Conference and Exposition* (PSCE06), Atlanta, pp. 2206-2211, October-November 2006.

[ROB 05] ROBERTS B., MCDOWALL J., "Commercial successes in power storage – advances in power electronics and battery applications yield new opportunities", *IEEE Power and Energy Magazine*, vol. 3, pp. 24-30, 2005.

[ROJ 03] ROJAS A. (Beacon Power), Integrating flywheel energy storage systems in wind power applications, http://www.beaconpower.com/products/EnergyStorageSystems/docs/Windpowe_2003.pdf, 2003, accessed October 2008.

[RTE] RTE, Référentiel Technique de RTE, http://www.rte-france.com/htm/fr/mediatheque/offre.jsp, accessed October 2008.

[RTE 04] RTE, Mémento de la sûreté du système électrique, Edition 2004, http://www.rte-france.com/htm/fr/activites/garant.jsp, accessed October 2008.

[RTE 09] RTE, Règles relatives à la Programmation, au Mécanisme d'Ajustement et au dispositif de Responsable d'Equilibre, Edition from March 3, 2009, http://www.rte-france.com/espace_clients/fr/visiteurs/offre/offre_marche_regles.jsp.

[SEI] SUSTAINABLE ENERGY IRELAND (SEI), VRB ESS Energy storage and the development of dispatchable wind turbine output: feasibility study for the implementation of an energy storage facility at Sorne Hill, http://www.sei.ie/index.asp?locID=99&docID=932, accessed October 2008.

[SOR 05] SORENSEN P. *et al.* (Risoe), Wind farm models and control strategies, technical report, available at: http://www.risoe.dk/Knowledge_base/publications/Reports/ris-r-1464.aspx, August 2005.

[UCT] UCTE, UCTE operation handbook, available at http://www.ucte.org, accessed October 2008.

[VRB 05] VRB Energy storage for voltage stabilization: testing and evaluation of the Pacificorp vanadium redox battery energy storage system at Castle Valley, Utah, EPRI, Palo Alto, CA., 1008434, 2005.

[VRB 07] VRB POWER SYSTEMS INC., The VRB Energy Storage System (VRB-ESSTM) – Use of the VRB Energy Storage System for Capital deferment, enhanced voltage control and power quality on a rural distribution feeder utility – a case study in utility network planning alternatives, March 2007, online at: http://www.vrbpower.com, accessed October 2008.

Chapter 2

Transport: Rail, Road, Plane, Ship

2.1. Introduction

The transport of goods and people has become more and more important in the modern world and, whatever the means used, whether rail, road, plane or ship, hydrocarbons play a major role, even though the situation varies according to country for rail transport. Electricity already plays an important role in the four types of transport, for improving the function of internal combustion engines (ICEs) or for replacing mechanical or hydraulic devices.

Hydrocarbons generate pollutants and greenhouse gases, despite great progress in mastering the function of ICEs. Two sources of energy are possible replacements for hydrocarbons: electricity and hydrogen. Both have the same handicap: they are difficult to store. In this chapter we are interested in devices that allow electrical energy to be stored despite the constraints of transport.

Electrical energy pervades all means of transport. It is often produced on vehicles, but their functioning requires it to be stored. We can classify vehicles into three categories as follows:

– those for which electrical energy is a secondary energy compared to the energy created by hydrocarbons and transformed by the ICE;

Chapter written by Jean-Marie KAUFFMANN.

– those for which electrical energy is the principal energy, or even the only energy, and in this case storage is a major limiting factor;

– those presenting a balance between the two forms of energy: hybrid vehicles.

The means of storage will be discussed elsewhere in this work, and we will also present the conditions for use and the constraints produced by the transport application and the levels of voltage that are required.

2.2. Electrical energy is a secondary energy

2.2.1. *Ground transport*

2.2.1.1. *Car with an ICE*

Electrical energy is an essential source of energy for the functioning of a car even with an ICE (petrol or diesel). The vital elements that use electrical energy are the ignition system of the ICE and the starter. In addition to these two basic elements there are functions that cannot be carried out using mechanical energy, such as lighting and driving of the windscreen wipers.

A natural and progressive evolution has taken place towards electrification of a certain number of functions – fan drives, direction assistance, the pump for assisted braking – owing to the flexibility of electrical energy versus mechanical or hydraulic energy, which has thus allowed new architectures in the interior of the vehicle and often a better efficiency. Currently, practically all the air-conditioning functions are electric: fan motors, the control of air valves and gates. The ignition control and injector control are electric and we are going towards valves (1 to 2 kW) with electric controllers (replacing the camshaft), a turbocompressor with electric assistance allowing improved combustion and reduction of polluting emissions. In this way, numerous functions that are generally carried out using mechanical energy will be electrified. Eventually, the current tendency is a move towards “drive by wire” where mechanical links will be reduced to a strict minimum. The increase in electrical power is estimated at 120 W per year.

We must not forget the other accessories for comfort that are appearing almost systematically on all vehicles: onboard radio, an information system for controlling correct functioning (onboard computer), various sensors, car phone, and even, to a lesser extent, GPS[1].

1 GPS: global positioning system.

The electrical architecture is completely modified and aside large consumers of electricity, such as the starter or the lighting, we find lower power circuits and especially small signal circuits for control. The wired links, the multiplexing, the VAN (Vehicle Area Network) bus, the CAN (Controller Area Network) bus, and the power bus are all located together; on the one hand, posing over-heating problems, and on the other hand, posing EMC[2] problems. Interference and dysfunction of the functionalities that have become vital, such as ABS[3], ESP[4], ASR[5], or speed regulation, must be avoided.

2.2.1.1.1. Generation of electrical energy

The source of electrical energy is an alternator driven by the ICE. The transition from a DC current commutator generator to a claw-pole alternator with a three-phase rectification by diodes has allowed the rotation speed to be increased and electrical energy to be supplied even at low speeds or when decelerating. The efficiency of these alternators is not very high (in the order of 50%). The tendency to increase the installed power has encouraged improvement of the design and realization of alternators in order to increase the specific power and efficiency. The largest alternators attain 2.5 kW with an efficiency of 85%.

2.2.1.1.2. Voltage level

For a long time, the installed power was weak, with a network of 6 V and a lead battery composed of three elements in series. We quickly progressed to a network of 12 V. Large sedans even need two separate batteries. The increase in power has led to interest in a network of 42 V, which would have favorable direct repercussions for drivers due to the resulting reduction in the mass of copper on board. We have come to an industrialization problem.

It is necessary to modify the design and characteristics of all the controllers, and to put in place a new fabrication chain after much testing. For the moment, we will not control the lighting circuits with this voltage, although some headlights, such as xenon lights, need voltages that are much higher (high voltage) than 12 V. A car that has a 42 volt network does not currently exist. The problem is different for hybrid vehicles, which require a direct current bus with a voltage level that is much higher, and the 12 and 42 V networks coexist on some hybrid models.

Classical vehicles all use lead batteries, which have progressed greatly in order to satisfy the peak power needs of the starter and the mean power needs associated

2 EMC: electromagnetic compatibility.

3 ABS: anti-lock breaking system.

4 ESP: electronic stability program.

5 ASR: acceleration skid control.

with switching on all the auxiliary functions. As all these auxiliary functions, especially those related to comfort, are not all used simultaneously, energy management allows reduction of the battery size after taking into account the power installed on the vehicle.

The capacity of a battery is expressed in amperehours (Ah)[6]. Manufacturers specify the capacity for a constant discharging current, which is equal to the nominal current. The energy and power specifications for batteries have increased as a result of several techniques such as:

– alveolate anodes;

– recombination of gases (the electrolyte moistens the separators, made of glass microfibers, which help to maintain the anodes); the casing is waterproof and has an explosion-proof valve (AGM[7] battery);

– the electrolyte is immobilized using a silica gel.

Maintenance is limited and the reliability is markedly improved.

The electrical energy of installed batteries varies between 40 and 70 Ah according to the size and the motorization the vehicle. Diesel engines require more powerful batteries for start-up.

2.2.1.2. *Buses and coaches*

Buses and coaches all have diesel engines. With regard to a tourist coach, in addition to the power of the ICE, it is the functions for comfort that require lots of power, considering the size of the coach: air conditioning, lighting, radio equipment, etc.

We must not forget the security systems: pneumatic braking with passive security which requires a compressor driven by an electric motor and also an eddy-current brake. A magnetic field is created by a direct current and as a result of the rotation, currents are induced in the rotating part linked to the wheels, generating losses and therefore dissipation of kinetic energy.

The supply network is generally at 24 V, with two 12 V batteries in series, and the stored energy reaches 300 Ah. An alternative network to the 24 V network is

---

6 Ah: amperehour – unit expressing the quantity of electricity in a battery. One amperehour is equal to 3,600 Coulombs. A battery with capacity of 40 Ah can deliver a nominal current of 5 A over 8 hours.

7 AGM: absorbed glass mat battery.

often created in order to be able to power systems for comfort, such as audiovisual equipment or security systems.

2.2.1.3. *Heavy goods vehicles and utility vehicles*

Commercial vehicles of less than 3.5 metric tons can be treated like tourist cars with the same level of voltage. Only heavy goods vehicles with particular functions merit separate consideration. For example, this is the case with refrigerated vehicles that need onboard power and, therefore, more significant energy storage. The necessary power is estimated at 1.5 kW for a small volume of refrigeration and so the installed power must be doubled. This means that firstly a more powerful alternator must be installed, and secondly, the size of the battery must be increased. Other solutions can be envisaged. We will cover these a little later when considering heavy goods vehicles.

For heavier vehicles the safety constraints are of the same nature as for buses (pneumatic braking, electromagnetic retarder). It is also necessary to consider the comfort of the driver, power steering, heating and air-conditioning of the cabin, even at standstill. The energy installed is in the order of 200 Ah or even more.

Energy storage becomes even more important for refrigerated lorries or for lorries undertaking extremely long distance journeys in Europe or America. The rules regarding resting time are very strict and so the lorries remain parked in rest areas for substantial durations. Leaving the thermal engine running is not an economically viable solution as the thermal engine efficiency equals only 9% to 11% when stationary.

One plan was to equip rest areas with electricity sockets to ensure energy supply to all needed elements. Other solutions include installation of an auxiliary generator on fifth wheel tractor or on the trailer. A small generator with power of the order of 7 kW meets the need. It only functions when the vehicle is stationary but it can work continuously on refrigerated lorries.

This auxiliary power solution (APU[8]) is a more general solution and leads to research works on use of fuel cells. It is also being considered for top of the range cars as the better fuel cells efficiency would allow them to save between 0.5 to 1 liter per 100 km, especially when air-conditioning is highly used.

The high temperature fuel cells (SOFC[9]) would be most suitable. SOFCs typically work between 750°C and 850°C. A reformer converts the used hydrocarbon into carbon monoxide and hydrogen, which is fuel for the SOFC, the

8 APU: auxiliary power unit.

9 SOFC: solid oxide fuel cell.

combustive being air. Much research has been done on these systems. Aside from problems concerning lifespan, thermal cycling, etc., related to fuel cells, a tricky problem relates to the reformer; at the moment it is in fact the quality of the fuel (petrol or diesel oil) which poses a real problem because it must be sulfur-free. It is clear that this solution will only be economically viable if the fuel for the main engine is the same as for the APU.

For vehicles with a specific purpose, such as packer body trucks or vehicles with an elevating tailgate, the power needed varies between 1.5 and 30 kW [MAR 07]. For the moment these functions, which are often electrified, are powered by the alternator and the ICE. In order to be able to use industrial engines, an alternative network of 400 V must be created knowing that the battery voltage level is at 24 V.

2.2.1.4. *Two wheels*

Motorbikes or scooters need only a small amount of electrical energy. The batteries are of a reduced size despite development of electric auxiliary functions, such starter assisted driving systems or control devices for the engine. The battery voltage has increased from 6 V to 12 V.

2.2.2. *Air transport*

Without doubt, electricity is increasingly essential in aviation. Flight control use mechanical and then hydraulic drives on the largest planes. Electrical energy is needed for navigation instruments, for the comfort of passengers, and at start-up for engines and turbines.

For some years now we have what is described as an "all-electric" plane, with installation of electro-hydraulic controls leading to a hybrid architecture that ensures redundancy. The electric power installed increases regularly as it goes from 120 kW for the A320 up to 500 kW for the A380 [VAN 08]. In the future, there are plans to require power of 1 MW or even more. The energy is supplied by a turbine driving a high speed generator.

The voltage levels, initially 115 V AC[10] 400 Hz, are diversified. The frequency is no longer constant, and we are going to different power buses at 115 V AC, 230 V AC, 270 V DC[11], and 28 V DC, interconnected by DC-DC or DC-AC converters. The latter source powers control, switchgear and controlgear. Onboard energy must ensure the essential functions of starting-up, control, and navigation.

---

10 AC: alternating current.

11 DC: direct current.

Distributed sources, such as fuel cells, are being studied. Again there is the problem of fuel as it is difficult to imagine carrying hydrogen on board, regardless of the storage method.

2.2.3. *Rail transport*

Rail transport may be partitioned into two classes: a first class, which can be qualified as being completely electric, powered by a catenary system or through a gliding contact on a third rail from a very powerful network, and a second class with a hybrid system where electrical energy is obtained via a diesel engine and a generator. Strictly speaking, only the latter class is the subject of this section. The majority of the energy is supplied by diesel, and aside from the 72 V network necessary for the equipment, there is no specific energy storage system.

2.2.4. *Maritime transport*

Cruise ships, ferries, or even cargo ships are often propelled by electricity which provides a smooth drive. This is particularly true for ships equipped with PODs: adjustable cradles with an electric motor that directly drives a propeller. The energy needs for ships are considerable; they have diesel engine generator sets connected to an onboard network. In addition, there is energy stored in batteries for safety and for the instruments and signals.

For boating like with yachts, the electrical energy needs are also significant. There is no difficulty at sea as electricity generation is ensured by a diesel engine generator. The restrictions are stronger when docked; these boats are often obliged to keep their generators running, generating pollution and noise. Here, again, research is being undertaken to find a more economic and less polluting solution, such as high-temperature fuel cells of SOFC type. The restrictions of thermal cycling are reduced as the generator functions continuously just like the fuel cells used during stationary supply.

2.3. Electrical energy: principal or unique source

Stored electrical energy as the only source of electrical energy has not really been developed in transport applications except for vehicles with dedicated circuits, for example, trams, trolleybuses, underground trains, trains with a fixed electric power infrastructure using catenaries and pantographs, or powered from the ground. Road vehicles or electric service vehicles are only used in fleets (at airports, factories, for deliveries in restricted zones, etc.) but the low dissemination leads to disproportionate costs considering economic and environmental advantages. Only a voluntary policy, which emphasizes how substantial gains can be made in battery performance, can lead to evolution of mentalities.

2.3.1. *Electric road vehicle*

The use of only electrical energy for transport applications has always been an objective for electricians and that would allow:

– use of clean energy, where there are no waste products and no pollutants;

– a homogeneous system across both power and command areas;

– energy recovery when braking.

Pure electric propulsion remains a long-term objective but the main stumbling block is still electrical energy storage. The competition with hydrocarbons is tough. As a rough guide, a liter of petrol supplies 40 MJ and the power to fill a reservoir is in the order of 10 MW. The energy density is extremely high and no other system can compete, as shown in Figure 2.1.

Also, it is difficult to imagine a cable carrying 10 MW of electric power and being as easy to manage. It would be necessary to greatly increase the voltage in order to reduce the cross-sectional area of the cable, leading to safety problems. Under 500 V, the current would be equal to 20,000 A; even if the energy storage system could absorb such a current, the cross-section of copper would be 2,000 mm^2 with a current density of 10 A/mm^2, which is already high and the diameter of the conductor would be 50 mm. Therefore, the mass of copper is almost 13 kg/m of conductor [RUF 07]. Moreover, losses would be without possible comparison with those at a petrol pump.

This handicap in storage and transferring energy is somewhat tempered by a greater efficiency of the electric motor compared to the internal combustion engine. We are going to review the constraints related to using electrical energy for the propulsion of a road vehicle.

The first electric car is very old, since the "Electrobat" was built in 1894 in Philadelphia. Peugeot's "Jamais Contente" reached more than 100 km/h in 1899. Developments stopped quite rapidly given the progress of the combustion engine and given the problem of energy storage in batteries. Production of electric vehicles stopped in 1918 and only restarted much later. The different vehicles proposed and available on the market were, therefore, conceived around the ICE. The ICE had been replaced by an electric motor.

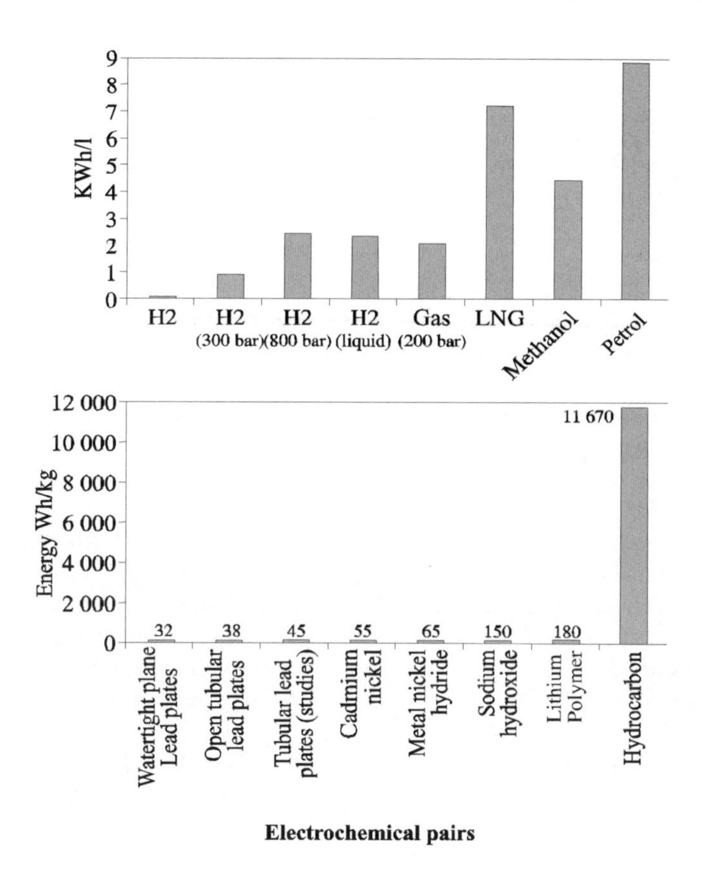

Figure 2.1. *Volumetric density of some fuels and comparison with batteries*

2.3.1.1. *Motorization*

Initially, direct current series motors with commutator were used, which enable variation of speed by acting on the voltage of the terminals. This kind of motor is well suited to electric traction (this type of motor was and is still used in electric locomotives, underground trains, trams, etc.). The torque is high at low speed and this solution enables direct drive without a gear box.

The solution used at the moment consists of using alternating current motors that are more robust, requiring less maintenance and delivering more power. They are combined with inverters that are supplied by a DC bus. The most robust motor is the induction motor (it has no rotating contacts). This engine is powered by a tri-phase inverter that regulates the voltage and the frequency of the power in such a way that the flux in the machine remains constant. Vector control techniques are quite suitable and enable efficient control of the speed and the torque. The overall efficiency is weaker as an asynchronous engine produces a torque only when there are losses in the rotor. The synchronous solution is preferable from this point of view; moreover, it permits the converter to function in better conditions if the field can be adjusted.

Electric motors are better suited to relatively high speeds whereas direct drive leads to speeds in the order of 400 tr/min at most. Variable reluctance machines may constitute an interesting solution. One of the first realizations was proposed by the Jarret brothers, but other structures are better suited. They often combine a rotor with many teeth with a stator using a multiphase coil supplied at a variable frequency, but they also have permanent magnets in order to improve performances. These motors have two major handicaps of heaviness and vibrations, even if they lend themselves easily to variable speeds.

The most elegant solution is to use a self synchronous motor, which has almost the same external characteristics as a direct current motor with a commutator and with independent excitation. In the brushless motor version, the inverter is a simple current commutator. The synchronous machine generally has permanent magnets in order to ensure its excitation; an inverted structure with an interior stator and an exterior rotor allows us to fix the motor directly on the rim of the wheel.

This technique was used by TM4 for motors with a power of 18.5 kW. ALSTOM installed asynchronous wheel motors on buses. This technique was also tested on a moving test bench (the ECCE[12] bench) but with motors of 30 kW. This solution may lead to new vehicle architecture saving space in the interior. There are three drawbacks: (i) the motors are mechanically independent and it is necessary to recreate the differential function using adapted set points for the inverters that supply the motor (electronic differential); (ii) the location of the inverters and their controller must be planned; (iii) the motors for the wheels increase the sprung mass and this can affect the road behavior of the vehicle. However, it is simple to install ABS, anti-slipping, energy recovery on braking and four-wheel drive if the four wheels are equipped.

The classical adopted solution consists of having a single electric motor linked, directly to the mechanical differential. Functions, such as ABS, are realized using conventional techniques. In all cases, it is important that the driver should feel the same sensations as for a traditional vehicle, especially with regard to the motor brake; it is obtained by an adapted control of the power electronics.

12 ECCE: test bench for components of an electric chain.

2.3.1.2. *Batteries*

Four types of batteries compete for use in electric vehicles[13] [THE 06]:

– technology using lead;

– technology using metallic nickel hydride;

– technology based on lithium (lithium ion or lithium polymer);

– technology using sodium-nickel chloride.

The mass of the electric vehicle is fundamental to the design. Batteries make a significant contribution to this mass. The data given in the literature are not always coherent but the trend is clear. In order to be able to drive autonomously for some 100 km, 30 kWh are required, which requires approximately 850 kg of lead batteries, or 270 kg of lithium-ion batteries. It is also necessary to house these batteries, and their parallelepiped form is not an advantage. Lithium-based or nickel hydride-based batteries are more flexible and can be more easily fitted to a vehicle. Therefore, a car must be designed from the beginning as an electric vehicle and this is the trend that is observed in vehicles that could be described as second generation.

There are multiple difficulties associated with batteries:

– Stored energy: lead batteries have a stored energy of 40 Wh/kg whereas new batteries based on lithium or nickel should be able to reach 220 Wh/kg and 100 Wh/kg, respectively.

– Cycling: the number of cycles that a battery can withstand without significant reduction of its capacity. Unlike a battery that is used in a vehicle with an internal combustion engine, batteries in electric vehicles must undergo large discharges, which reduce their lifetime. Lead batteries have a low capacity for cycling (180) and this figure is only slightly better for batteries based on lithium (1,000) or nickel (1,000), where high-energy batteries are being considered. The capacity for cycling is greater for high-power batteries (the data becomes 1,000, 200,000 and 250,000 respectively) but these are not suited to electric vehicles.

– Autodischarge: nickel batteries are penalized after hydrogen diffuses across the electrolyte.

– Prices: lithium batteries are very expensive compared to lead batteries. NiMH batteries are a little less costly.

– Behavior at low temperatures: the performance of lead batteries is reduced at low temperatures; this is the same for other types of battery, especially for lithium-ion technologies. This handicap is important as an electric vehicle should be able to start-up when the temperature is –20°C.

13 Nickel-cadmium batteries will no longer be authorized due to the toxicity of cadmium.

– Recycling: lead batteries can be recycled as ensured by their manufacturers. Methods to recycle other types of batteries should be developed at the industrial level by manufacturers, given the quantities of such batteries used in electric vehicles.

2.3.1.3. *Battery efficiency*

All the energy of a battery cannot be recovered due to different losses. We distinguish faradic efficiency, whose value is tending towards 1 for new batteries (NiMH or lithium) while it is less for lead batteries, the electric efficiency linked to losses due to the Joule effect, and the efficiency of the battery charger linked to the power electronics. It is important to keep account of the global efficiency which varies from 70% to 90% for batteries to get a sufficient storage capacity for the use of electric vehicles.

2.3.1.4. *Voltage level*

The voltage level in electric vehicles is not normalized and depends on an optimization carried out by each manufacturer, which considers motorization choices and also the mass and volume constraints. The voltage level of the DC bus is often in the order of 300 V, whereas it was around 100 V for the first designs. This leads to having to connect many batteries in series. Recharging to the maximum level should be done regularly in order to balance the voltages between the different elements.

2.3.1.5. *Battery characterization*

To recap: a battery is characterized by its specified energy and its specified power, the domain in which it functions, its capacity to withstand charging and discharging (cycling), its behavior at low and high temperatures, autodischarge, aging, lifetime, price, etc.

In an electric vehicle, the battery is the only source of energy. For a traditional vehicle, the petrol or diesel gauge gives a good indication of the amount of fuel left and the vehicle's computer can calculate the range of the vehicle. It should be the same for an electric vehicle. Therefore, it is necessary to monitor the state of the charge (SOC[14]) in order to be able to correctly determine the range of the vehicle. Measuring the state of charge is not at all easy for lead batteries. Techniques have been developed for NiMH and lithium batteries. Given a known state of charge, the current would have to be integrated so as to deduce the electric charges removed from or added to the battery. The theoretical specified energy is not completely available except when the current drawn by the generator is infinitesimal for a very

14 SOC: state of charge.

long time. As soon as the current density is non-negligible and the system is supplying power, the potential difference is reduced, due to polarization phenomena and ohmic drops resulting from the resistance of the electrolyte and the resistance of its constituents. Therefore, only a no-load signal gives precise data on the state of charge of batteries and the battery must be rested for some time in order to replenish the charge. It must not be forgotten that the state of charge depends on the temperature of the battery [ELK 07].

An indicator for an electric vehicle is the DOD[15], which describes the rate of discharge permissible without leading to irreversible loss of energy in the battery. For electric traction, it is best to choose batteries with a large DOD (for example 80%).

We also introduce a relatively subjective indicator, the SOH[16], which takes into account the progressive damaging of the battery, its capacity to be charged, internal resistance, voltage and current during autodischarge. Therefore, the SOH is just as important as the SOC for an electric vehicle [COX 00].

2.3.1.6. *Auxiliary functions of an electric vehicle*

In addition to traction, it is also necessary to satisfy auxiliary functions. Having seen their progressive electrification in a traditional petrol or diesel powered car, this problem is solved in a similar fashion except for heating and air-conditioning. There is no longer a source of heat because even if the inverters, and sometimes the motors, have to be cooled, the temperature levels are not at all in the same order of magnitude of those of a traditional radiator.

It is difficult to imagine a small burner functioning on hydrocarbon fuel in order to ensure heating, and it is a pity to use so much electric energy in supplying a resistance. Suggested solutions include thermoelectric refrigeration (the Peltier effect) or magnetocaloric[17] refrigeration, but this last technique is still undergoing research and development. Auxiliary functions are major consumers of energy and improving their efficiency would allow an increase in the range of the vehicle.

Refrigeration is not only needed for the vehicle cabin, but also for the components required for propulsion, for example, the power circuits, or the motor,

15 DOD: depth of discharge.

16 SOH: state of health.

17 Depending on the direction of variation of magnetic induction, a material absorbs or supplies heat energy. Specific materials are needed, such as gadolinium, in order to have a significant effect on the ambient temperature.

and batteries or ultracapacitors[18], which have relatively narrow temperature zones in which they can operate (temperatures less than 60°C). Other batteries must be maintained between 60°C and 80°C in order to perform well (lithium polymer batteries, for example).

2.3.1.7. *Battery recharge*

The problem of charging a battery is an important problem. The charger may be exterior or may be integrated on the vehicle to enable recharging from a traditional alternative network. Charging by induction without physical contact was envisaged, but the efficiency of the system is not particularly impressive even at high frequency, due to significant interference between the two elements coupled magnetically to ensure energy exchange.

As a rough guide, we present the recharging conditions for a lead battery of 96 Ah, as well as the discharge limits given by the manufacturer [ELK 06]:

– capacity to discharge in 5 hours: 98 Ah;

– range at a current of 200 A: 14 min;

– normal recharge current: 19.6 A;

– rapid recharge current: 39.6 A;

– maximum transitory recharge current: 100 A;

– maximum current for discharging in 1 minute: 450 A.

Similar reports are valid for other types of battery.

Rapid recharge is conceivable, generally outside of the usual garage for the vehicle; the installed power must be significant and this requires an important investment and a commercial management system, aside from the problem of normalization. The power installed at a usual garage can be markedly weaker since the charge is normal. In all cases rapid connections, which have been secured and standardized, are required. Moreover, the charger should be suited to the type of battery and only the solution of a charger integrated within the vehicle seems foreseeable. The mass of the vehicle is increased by that much.

It is possible to make a standard exchange: a charged battery swapped for a flat battery. This assumes a specific infrastructure, a rapid connection system, and a significant stock of batteries. At the moment, this is only conceivable for fleets.

---

18 Ultracapacitors function in the same way as capacitors but use the dual-layer phenomenon enabling a very small distance between positive and negative charges. The principle behind their fabrication also enables the active surface to be increased. Currently, ultracapacitors of 10,000 F are manufactured but the voltage at the terminals does not exceed 2.5 V.

2.3.1.8. *Range extender*

To compensate for the loss of range in electric vehicles, a small generator is often installed, serving to recharge batteries when the state of charge is too low. The ICE functions at a stabilized speed, corresponding to a very good efficiency. There is no impact on the functioning of an electric vehicle (other than the fact that the voltage of the DC bus is necessarily more elevated because the direction of the current in the battery is reversed). The generator supplies some of the power to the propulsion engine. The schema is very close to that of a series hybrid. Recharging from the network is sometimes known as *plug-in*.

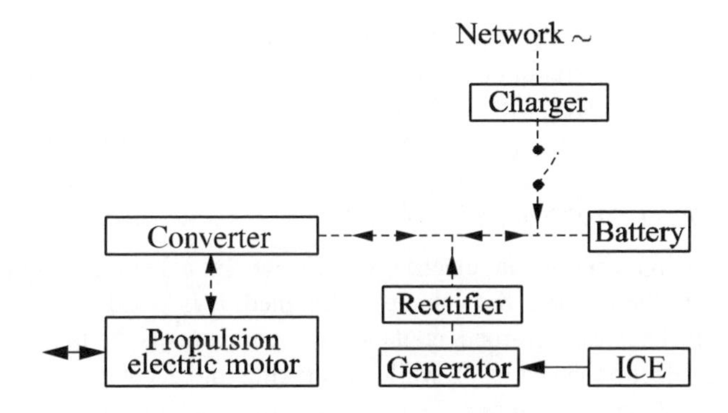

Figure 2.2. *Schema of a vehicle with range extender*

Fuel cells, generally PEFCs[19], can also be used as range extenders. Axane proposes AUXIPAC and Peugeot has presented concept coaches that use that specification (H_2O car).

2.3.1.9. *Examples of second-generation electric vehicles*

Electric vehicles that are or soon will be on the market (CleanNova and Blue Car in France) aim to use lithium-ion or lithium-polymer batteries so they have a specified energy that is distinctly higher and a shorter charging time compared with lead batteries, without any deterioration of the battery. Their characteristics are summarized next.

CleanNova is manufactured by SVE (Dassault-Heuliez). It has a motor with integrated differential TM4 supplied under variable voltage, lithium-ion battery of 16 to 30 kWh, rechargeable in 8 hours (16 A), 4 hours (32 A), 30 minutes (150 A); heating uses a heat pump. In the version with a range extender, the generator may be

19 PEFC: proton exchange fuel cell.

decoupled from its driving engine and coupled with the traction engine in order to function as a motor and supply a torque.

BlueCar is manufactured by Batscap. It has a motor of 30 kW at 10,000 tr/min, lithium metal polymer battery of 27 kWh, rechargeable in 6 hours, with a suggested lifespan of 10 years or 150,000 km, DC bus voltage of 243-375 V, range of 200-250 km, battery weight 200 kg, maximum speed 125 km/h.

For information only, Chevrolet announced a range of 60 km for a lithium-ion battery weight of 180 kg.

We can also mention recreational vehicles used for golf, for example, or four-wheeled vehicles that are light or heavy but limited in speed. Four-wheel drive vehicles are also produced with 12 batteries of 48 V and 240 A, permitting a range of 100 km [TEN]. Given the cost of replacement, these use lead batteries.

2.3.1.10. *Energy management and modeling*

Energy management in an electric vehicle, as in a hybrid vehicle, must be considered from the moment the vehicle is designed. It is based on the modeling of the vehicle (mechanics, tire-road contact), the components in the traction chain (motor, power electronics, battery), while also taking into account essential auxiliary functions or functions for comfort. The energy is managed by an onboard computer, which takes into account the SOC and the instructions of the driver. It is even possible to imagine a more intelligent system taking into account the profile of the actual journey (departure-arrival and itinerary defined by means of a GPS system).

2.3.2. *Heavy goods vehicles and buses*

There are not many examples of electric heavy goods vehicles. We cite trials with garbage collection trucks in Bordeaux and also electric or hybrid lorries manufactured by PVI (Ponticelli Véhicules Industriels) in collaboration with Renault Trucks. These trucks are used in the city for deliveries and for collecting household rubbish. They use lead batteries of 160 Ah under a total voltage of 456 V. The driving motor coupled with a gearbox has a nominal power of 90 kW. The range is 55 km.

Examples of buses also exist. GEPEBUS (a subsidiary of Gruau and Ponticelli) has proposed two models with 22 and 25 places, respectively. IRISBUS uses an electric motor of 140 kW for Europolis, and Zebra[20] batteries, allowing a range of 120 km.

---

20 Zebra: sodium-nickel chloride battery functioning at 300°C.

2.3.3. *Two-wheeled vehicles*

We have discussed four-wheeled vehicles but the problems are of a similar nature for two-wheeled vehicles, such as scooters or bicycles with assisted pedaling. Electric scooters are not very developed in Europe, whereas they are in Asian countries. A French manufacturer has stopped production due to an insufficient market but also due to a design that was suited to a scooter with an ICE. The lack of space is such that the allowable onboard weight for batteries is limited and the voltage level is in the order of 36 V. The power of the propulsion motor is around 2 kW.

An electric scooter can be redesigned using the technique of wheel motors and by linking the NiMH batteries with ultracapacitors that are better suited for energy recovery. Integrating the charger can pose a problem due to space and removing the batteries in order to charge them is more feasible. That requires a reliable and secure connection system. The batteries have an average capacity of 20 Ah.

Bicycles with assisted pedaling[21] all use electrical energy and all motorization techniques exist: engine in the front or back wheel hub, action on the pedals, roller rubbing on the tire. The range is directly related to the energy stored in the battery even if there is still a possibility of traveling by pedaling. The power of the auxiliary motor varies between around 150 W and 250 W and a well-trained cyclist can provide 250 W over 1 hour. The battery is, for example, the NiMH type, at 36 V and 7.5 Ah. This enables a direct power supply to the motor without raising voltage. The battery is generally removable to ensure rechargeability.

a) Wheel motor 30kW
nominal couple 700Nm
Maximum couple 6,000Nm

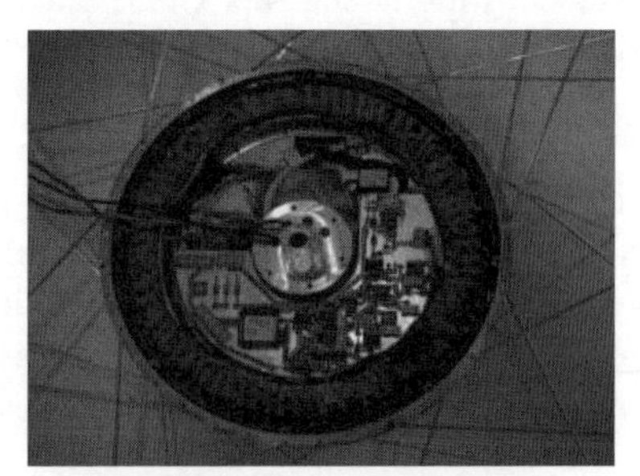

b) Engine for assisted pedaling
placed in the hub with
integrated power circuits

Figure 2.3. *Example of wheel motors (photography: University of Franche-Comté)*

21 The cyclist must provide some of the energy so that this type of bicycle is not classified in the motorcycle category.

We could equally describe recreational two-wheeled vehicles, such as electric scooters, or more elaborate ones, such as the SEGWAY.

2.3.4. *Guided vehicles (locomotives, underground, tram, trolleybus)*

The powering of all guided vehicles, such as locomotives, TGVs, trams or trolleybuses, do not pose any problems of energy storage for their propulsion. The powering is done using catenaries and pantographs on the railway except in some countries where there is still a third rail and power is supplied by gliding "shoes". It is the same in the case of underground trains. There are more diverse solutions for trams, which are conceived to circulate in the city and have a guiding circuit.

For very touristic zones the solutions that do not involve catenaries are being researched. Powering from the ground is a solution that has been used but there is also research being done into storing energy on board the vehicle. Two means of storage are favored at the moment: inertial storage[22] and ultracapacitors. Recharging stations for the inertial storage at each stop could be used but that poses the problem of transferring a large quantity of energy in a small amount of time. As an example, an inertial storage system used for a tram can provide 3 kWh of energy and 300 kW of power. An older application, such as the electrogyrobus, brings together inertial storage of 3.3 kWh and banks of capacitors.

Storage, whether it is in kinetic, electric, or chemical form, enables recovery of energy when braking and limits the total consumption of energy on a route, but also enables boosting at start-up. Gains reach 15% and management is facilitated as the profile is perfectly known.

Storage of electrical energy is essential for auxiliary functions when there is a defect in the traditional power supply (ripped catenary for example). During an unexpected stop, the internal signals of the train, underground or tram must be powered, and the lighting and air-conditioning must be ensured as much as possible. Batteries are put in the passenger carriages. Locomotives also use energy storage by means of a battery of 72 V. Lead batteries are used.

In the end, we imagine solutions permitting provision of more power, such as, for example, a fuel cell generator, but then it is necessary to install stored hydrogen or a hydrocarbon reformer in the case of a PEFC cell, or to use an APU of the SOFC type. In the latter case, the energy source is a hydrocarbon such as petrol or oil.

---

22 Energy is stored as kinetic energy in flywheels turning at great speed. It is released when needed, leading to a reduction in the speed of the flywheel.

2.3.5. *River transport – yachts*

To improve passenger comfort (reduce noise and pollution), it is desirable to replace the ICE with an electric motor for boats based in canals or lakes. There are not many constraints on the start-up couple, and the speed of the boat is slow in order to allow passengers to make the most of the view. Therefore, the motors are not very powerful and the range depends on the quantity of batteries that can be brought on board. There is space available in the hold. Recharging can be done at the embankment at nights; however, ventilation of the hold should be maintained during recharging. Such boats exist, for example, on the Saint Martin canal in Paris or on the Doubs for going to Doubs fall.

Other applications exist for little pleasure boats either for powering a small auxiliary motor or to provide energy for the navigation instruments.

2.4. Electrical energy complementing another source – hybridization

The narrow range of electric vehicles due to energy storage problems has led to dual-mode vehicles, which generally bring together an ICE with an electric motor, but one can equally bring together two different sources of energy, for example, chemical and electric. Clearly the two modes can be combined in many different ratios, which has led to commercial names that sometimes need deciphering, for example, *stop and go*, *mild hybridization*, *full hybridization*, *boost*, *downsizing*, etc. Different architectures are distinguished according to the coupling between internal combustion and electric motorizations:

- parallel architecture;
- series architecture;
- road coupling.

In all cases, the objective is to limit the consumption of hydrocarbons and, therefore, to limit pollution and the emission of greenhouse gases, indeed to be able to function in urban mode with zero emissions. The range depends on the capacity for storing electrical energy. All architectures do not permit the same energy saving functionalities. Some are more efficient but are restrictive as to the nature of energy sources.

The architecture and the level of hybridization are independent of the nature of the vehicle, whether it is a car, a truck, a building site vehicle, or a bus.

Hybridization of two-wheeled vehicles is not under consideration because hybridization always leads to an increase in the weight and volume, required in order to accommodate the thermal and electric components. The only evolution possible would be *stop and go*, which enables the ICE to cut out when stopped.

2.4.1. *Parallel architecture*

The most basic diagram is given in Figure 2.4, which illustrates what is known as "*stop and go*".

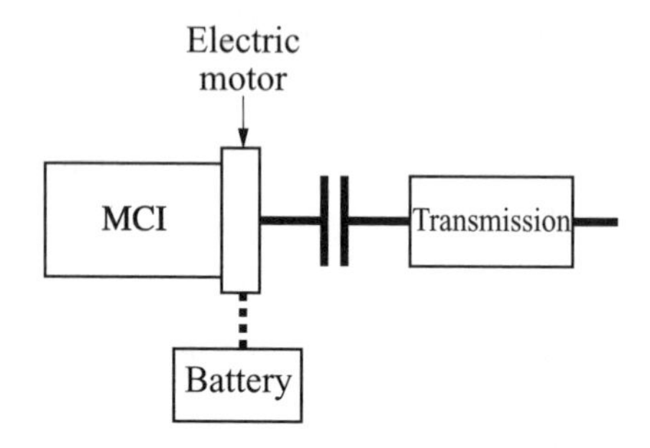

Figure 2.4. *Diagram of start and stop*

The electric motor can be placed directly on the shaft of the heat engine, as shown here, or connected to the mechanical system via pulleys and belts. The electric machine has two functions: charging the batteries and starting up the car, or even acting as the driving engine when zero emissions are required. Performance depends on the power of the motor and the size of the battery. Gains in consumption are very limited (around 5%) and only occur in urban mode. Saving energy when braking is hardly possible.

The provision of electrical energy enables the size of the ICE to be reduced and, therefore, reduces friction and reduces consumption (*downsizing*). In order to improve performance in cases where power is required, for example, during acceleration, the electric motor brings a supplementary type and this allows the engine to be boosted.

Parallel architecture is quite a bit more complicated, as shown in Figure 2.5. Parallel architecture is based on mechanical coupling, which is generally realized with epicycloidal trains enabling torques to be added despite having different rotation speeds from the three shafts. Engaging the clutches when necessary allows decoupling and enables the vehicle to function as a traditional vehicle or as an electric vehicle with zero emissions. Arrows indicate the possible directions of energy flow. The ICE can only supply mechanical energy, be it to the wheels or else to the generator, which serves to charge the battery or supply the auxiliary functions, or else via the mechanical coupling to the electric engine which then functions as a generator. It is this last path that enables energy saving during braking.

In this system, all the energy comes from the ICE and the saving in consumption is a result of good energy management thanks to the different converters each functioning at their optimal points. Adding a battery charger that would be supplied by the power grid would diversify the sources of energy.

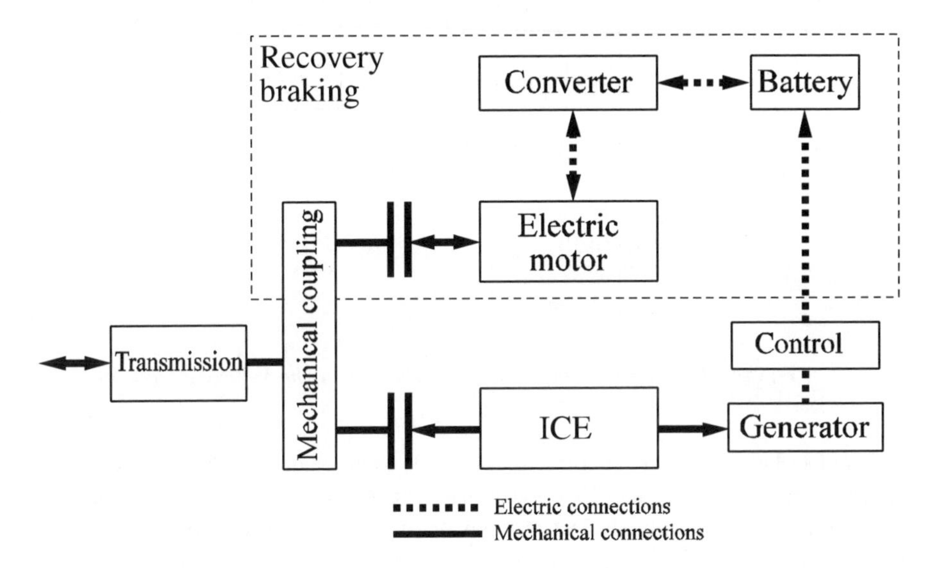

Figure 2.5. *Parallel architecture*

This architecture was first developed by Toyota for the Prius. Other manufacturers launched themselves along this path, making it more complicated with one or two electric motors narrowly inserted in the mechanical coupling and with connections enabling the different contributions to be controlled.

The increase in the weight of the vehicle, as related to the mechanical coupling, the electrical motors, the converters, and the batteries, should be noted. The only possible gain is in reduction of the size of the ICE. Therefore, it is not surprising that, for the moment, hybrid technology is limited to top of the range vehicles.

Standard levels for the voltage of the DC bus do not yet exist. It would be advantageous to choose a high voltage in order to reduce the diameter of conductors in electrical machines. This high level penalizes the batteries. If a 540 V DC bus is required, corresponding to a 400 V rectified alternating network, it would be necessary to put 150 elementary lithium ion or NiMH batteries in series. This poses problems of balance in charging, and also during functioning since it would be desirable for all the batteries to work in the same conditions. Therefore, the trend is turning towards having three or four different voltage levels in the same vehicle. The diagram in figure 2.6 is inspired by the electrical architecture of the Lexus (Toyota).

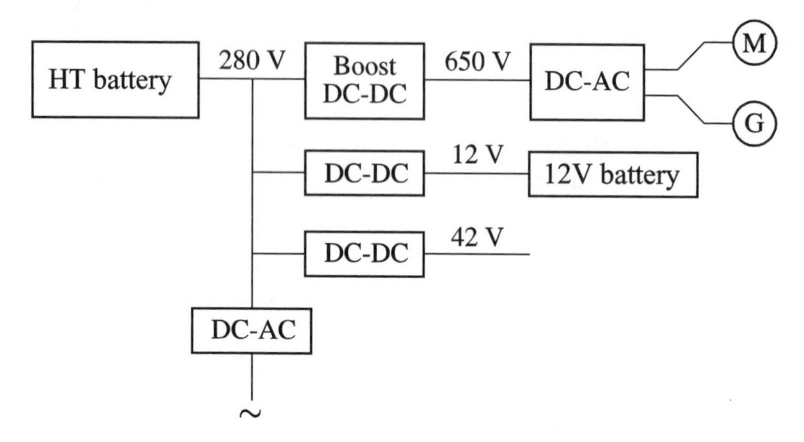

Figure 2.6. *Diagram of electrics for a vehicle with four voltage levels*

The DC-DC converters enable voltage levels to be changed. The motors and the generator are alternating tri-phase machines. The AC level enables equipment that was designed for traditional electrical power distribution system to be plugged in and, therefore, specific equipment does not have to be developed, for example, for air-conditioning. The complexity is high for both electric and mechanical components. A supervisor unit for energy management needs to be added, so that the charge levels in the batteries will enable energy saving during braking (the battery must not be fully charged since in that case it cannot accept any more coulombs of charge).

Thanks to energy saving during braking, consumption is reduced compared to a traditional vehicle in urban traffic or on the road. It is unchanged or even increased on the motorway due to the increase in weight.

2.4.2. *Series architecture*

A diesel electric locomotive typically has a series structure, but for a road vehicle, the architecture is complicated by adding storage elements such as the batteries, ultracapacitors, or an inertial fly-wheel. The general diagram is shown in Figure 2.7.

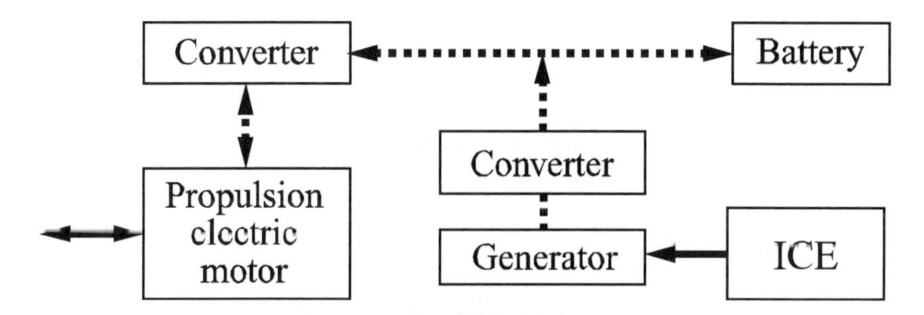

Figure 2.7. *Classical series architecture*

The ICE drives a generator (generally an alternator). The alternating voltage is rectified and adapted to the voltage of the battery, which fixes the voltage of the DC bus. Then, the classical chain of propulsion for an electric vehicle is used, with either a central motor and a differential, or a more distributed architecture with several motors, even wheel motors.

Arrows in the figure show the direction of energy flow. The ICE can only provide mechanical energy. The converter linked to the generator is unidirectional. Energy saving while braking is carried out by the electric motors for propulsion and their associated converters. The energy is stored in the battery, and therefore, the state of charge of the battery must be monitored so that it can effectively store this energy.

As the different energy converters have been placed in series (converting mechanical to electrical and then to mechanical energy again), the global efficiency is equal to the product of the efficiencies and is structurally weak even when the ICE is functioning under optimal conditions. The energy manager's role is to maintain a balance between energy provided by the ICE and that provided by the battery.

The series architecture is the only one that can be used for other types of generator, such as electrochemical generators or those using fuel cells. The diagram in the figure below shows such a case where a second type of transient storage (ultracapacitors) has been added for energy saving when braking.

The fuel cell is powered either by hydrogen or by hydrocarbons via a reformer. The word generator describes the fuel cell stack with all the auxiliary functions necessary in order to work, compressor and air humidifier, fuel treatment, refrigeration, controller for flows, and recirculation pump for hydrogen.

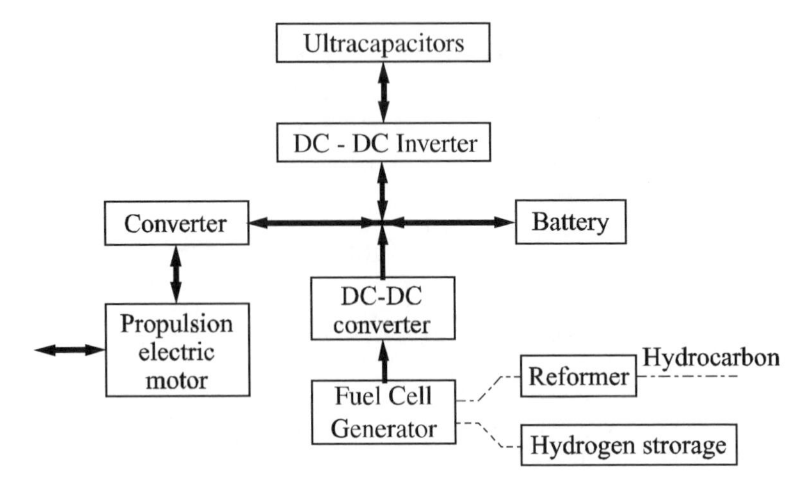

Figure 2.8. *Series architecture with fuel cells and ultracapacitors*

A fuel cell generator is unidirectional and provides direct current. Its global efficiency is now in the order of 35%, but it is hoped that this value is increased by the double effect on the efficiency of the electrochemical conversion and especially on the efficiency of the auxiliary functions to reach 55% to 60%.

The ECCE Bank [ELK 07] uses a series architecture. It enables testing of the behavior of different generators (variable or fixed speed generator, fuel cell generator PEFC type SPACT80 program), different electric motorizations, and energy storage by batteries, ultracapacitors, and inertial alternators.

2.4.3. *Coupling by road*

In the two architectures described in the preceding paragraphs, the ICE and the electric motor(s) are strongly coupled either mechanically or via an electric link. A third system comprises two propulsion systems that are independent, one acting on the front wheels and the other on the back wheels. The energy manager would rebalance the contributions of each. Therefore, we have would have a true electric vehicle, but using the mains to recharge the batteries and with energy saving on braking. However, this would be a traditional vehicle with a petrol or diesel engine with its characteristics. The performance of the two systems is adapted to specifications: zero emission functioning in urban mode and use of the ICE in outer-urban mode. The combination of the two modes permits improvements of the dynamic performance of the vehicle. Plug-in mode is essential for electric motorization.

Such a vehicle still does not exist commercially, but some research has been done in this area.

2.4.4. *Hybrid railway vehicle*

The power required of a locomotive is especially high at start-up in order to unstick the train and make it reach its cruise speed. Provided enough stored energy is available, hybridization is a solution. The basic structure is already electric as traction engines are direct current engines, induction motors, or synchronous engines with electronic control. In view of the lifespans required for railway components (more than 50,000 hours or 15 years) electrochemical solutions (batteries) present a big problem. The lifespan of ultracapacitors is not yet sufficient; inertial storage gives the best performance. Electromagnetic storage, with superconducting coils has not yet been used for such applications.

Manufacturers of railway components have carried out studies on hybrid locomotives. Japan, California, and Canada (Green Goat©) have announced the use of such locomotives. The diesel engine of the Green Goat is weakly powered (165 kW) and only serves to recharge batteries. The voltage level is 600 V. A saving of 60% in consumption has been announced, with an equivalent reduction in emission of greenhouse gases.

The SNCF is testing hybridization in the frame of the PLATHEE program, on a shunting locomotive (LHYDIE) with different storage systems: batteries, ultracapacitors, and inertial fly-wheel. Energy is supplied by a low power diesel ICE. The fuel cell solution has also been tested (SPACT80 program) with hydrogen storage. The voltage level is 540 V, like the ECCE bank, but with the target to increase to 750 V, which is the voltage level for trams.

2.5. Conclusion

The preferred system for electrical energy storage for transport applications is still electrochemical storage, i.e. batteries. Inertial storage is reserved for heavy systems, such as rail and especially trams, and maybe public transport. Auxiliary sources of power, such as fuel cells, are still heavily penalized by cost but also because they must be powered via a reformer or using hydrogen. Ultracapacitors are suitable for a pulsing source during charge and discharge and therefore are suited to energy saving during braking or to provide an extra torque at start-up or for selective needs (mounting the pavement for example).

The needs of road vehicles are summarized in Table 2.1 [KOH 07].

Application	Electric range	Minimal energy	Minimal power
Electric vehicle	150 km	20 kWh	40 kW
Plug-in HEV	Limited	10 kWh	40 kW
Hybrid bus	Limited	10 kWh	80 kW
Full hybrid	No range	1 to 3 kWh	25 to 50 kW
Mild hybrid	No range	0.5 to 1 kWh	5 to 20 kW

Table 2.1. *Energy needs of several road vehicles*

This table shows the importance of two aspects, energy and power – hence the decision to sometimes link two types of storage. The Ragone diagram shows the relationship between different storage systems.

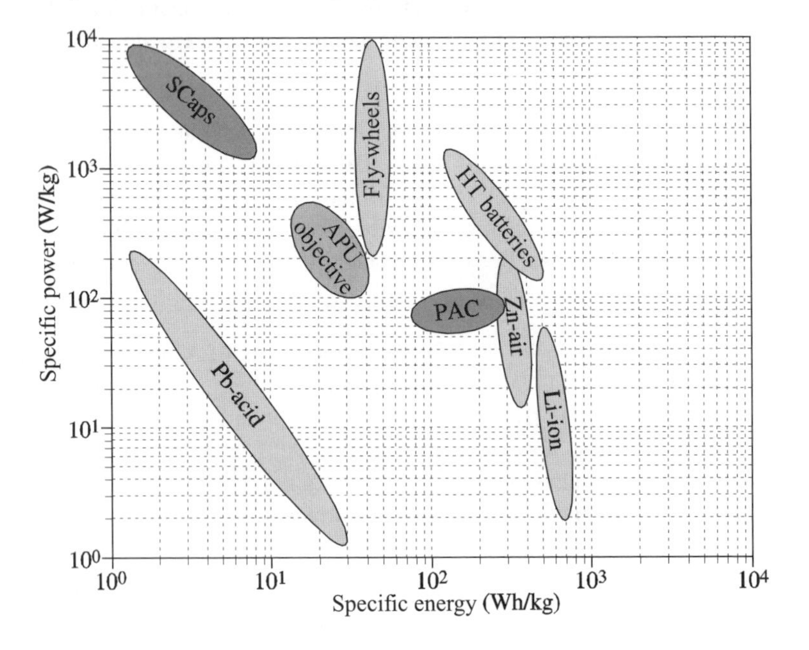

Figure 2.9. *Ragone diagram (current electrical energy storage systems)*

The different types of battery are not yet at the same stage of development. Lead battery performance is limited in terms of energy density and of lifespan, especially for electric vehicles that have a high DOD. They have the significant advantage of not being too expensive compared with the new NiMH, lithium-ion, or lithium metal polymer types. The performance of these new batteries is encouraging for both of the cited criteria. Progress has been made to minimize the influence of temperature (starting up cold) and to increase the lifespan of high-energy batteries.

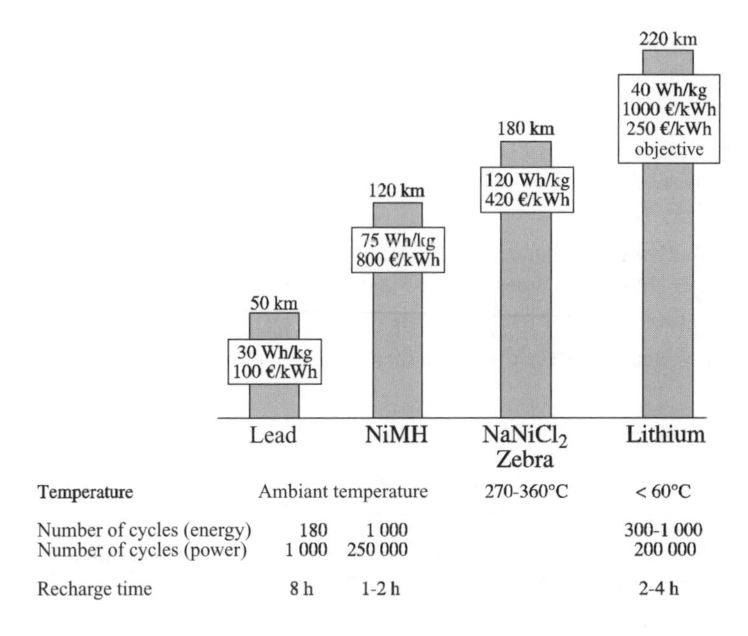

Figure 2.10. *Comparison of performances and conditions of use*

Finally, we have not taken into account the development costs of different energy storage systems (WTW[23]) nor the costs for their utilization (TTW[24]).

2.6. Bibliography

[BER 05] BERETTA J., BLEJ C., BADIN F., ALLEAU T., *Les Véhicules à Traction Électrique, le Génie Électrique Automobile, la Traction Électrique*, J. Beretta (ed.), Hermès, Paris, 2005.

[COX 00] COX D.C., PEREZ-KITE R., "Battery state of health monitoring combining conductance technology with other measurement parameters for real-time battery performance analysis", *IEEE 2000 19-2*, pp. 342-247, 2000.

[ELK 06] EL KADRI K., Contribution à la conception d'un générateur hybride d'énergie électrique pour véhicule: Modélisation, simulation, dimensionnement, PhD thesis, University of Franche-Comté, 2006.

[ELK 07] EL KADRI K., BERTHON A., KAUFFMANN J.M., AMIET M., Simulation and test of a modular platform hybrid vehicle, *Proceedings of Hybrid Vehicles and Energy Management*, Braunschweig, February 14-15, 2007.

---

23 WTW: well to wheel.

24 TTW: tank to wheel.

[GUA 04] GUALOUS H., HAREL F., HISSEL D., KAUFFMANN J.M., "Etude et réalisation d'une alimentation auxiliaire de puissance (APU) associant pile à combustible et supercondensateurs", *REE*, no. 8, pp. 90-100, September 2004.

[KOH 07] KÖHLER U., LISKA J.L., "Battery systems for hybrid electric vehicles – status and persperstives", *Proceedings of Hybrid Vehicles and Energy Management*, Braunschweig, February 14-15, 2007.

[MAR 07] MARTIN B., "L'intérêt de l'hybridation pour les véhicules industriels", *Journées thématiques DGA-CNRS Stockage de l'énergie*, ISL Saint Louis, October 23-24, 2007.

[RUF 07] RUFER A., "Le vecteur électricité, acteur majeur des systèmes énergétiques", *Journées thématiques DGA-CNRS Stockage de l'énergie*, ISL Saint Louis, October 23-24, 2007.

[TEN] VOLTEIS, Electric Car, 07430 DAVEZIEUX, www.tender.fr.

[THE 06] THEYS B., Les batteries pour le stockage de l'électricité dans les véhicules tout électrique ou hybride, Rapport Prédit III, February 2006.

[VAN 08] VAN DEN BOSSCHE D., "Des commandes de vol plus électriques: pourquoi, comment, perspectives", *REE*, no. 4, pp. 47-52, April 2008.

Chapter 3

Energy Storage in Photovoltaic Systems

3.1. Introduction

Stand alone or grid connected photovoltaic systems need storage in order to overcome the intermittence of their production. Stand alone systems are mainly concerned with batteries, whereas systems connected to the network have a wide range of applications, and as a result, numerous modes of storage are possible.

This chapter considers the issues for each of these categories of system as well as the storage technologies used.

3.2. Stand alone photovoltaic systems

3.2.1. *Principles*

The first category of solar photovoltaic systems is not connected to the electric grid, and is known as autonomous systems.

These systems are composed of photovoltaic modules (or solar panels), a battery and finally a regulator, enabling the management of energy between the module, the battery and the charge consumer. This connection is direct when the charge is powered by a continuous current. The system will eventually be composed of a

---

Chapter written by Florence MATTERA.

converter (or a DC/AC inverter), if the charges function with an alternating current (Figure 3.1).

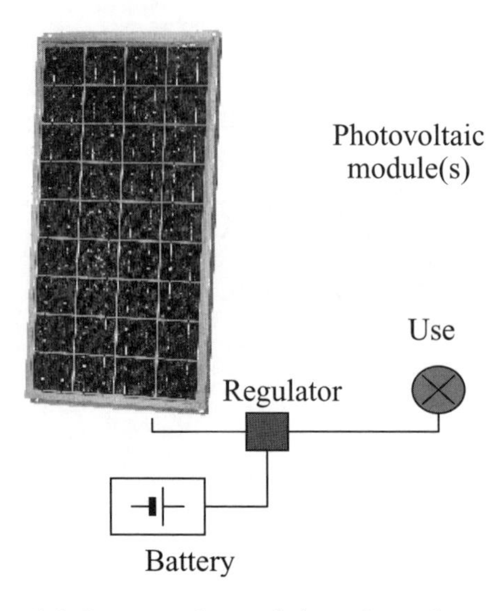

Figure 3.1. *Diagram of a stand alone photovoltaic system*

These systems have been in the majority for some 20 years, being set up during rural electrification campaigns in developing countries. For 2 billion individuals currently without electricity, often living in zones where there is a high amount of solar energy available, photovoltaic systems are one of the only energy solutions available.

3.2.2. *Indispensible tool: storage*

Autonomous photovoltaic systems need electricity storage in order to ensure a quasi permanent energy supply no matter what the varying level of sunshine. Depending on the applications, this energy supply will be ensured, for example, for 2 or 3 days for some weakly powered domestic systems to around a fortnight for professional applications, such as lighthouses or telecommunications relays. Figure 3.2 shows two examples of production and consumption over 1 day for two categories of autonomous systems:

– a weakly powered domestic system, or lighting "kit";

– a hybrid system (photovoltaic panels coupled with a small generator).

In the first case, production takes place mainly during the day and consumption takes place in an irregular way during the morning and then in the evening for lighting and eventually the television. In the second system, where the range of power is much greater, the diesel generator ensures additional supply in the evening and consumption takes place throughout the day due to additional electrical equipment being used.

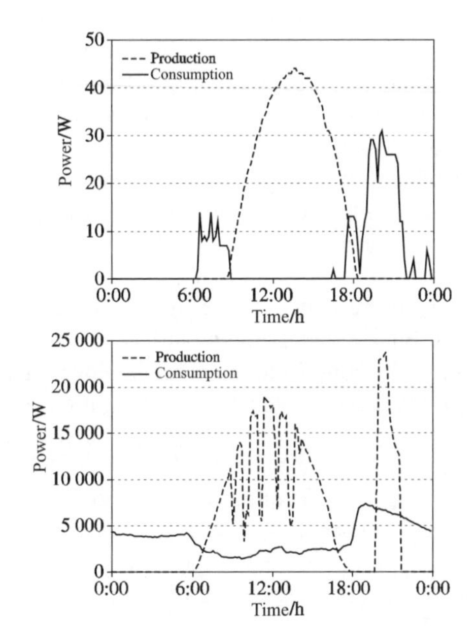

Figure 3.2. *Two examples of energy production and consumption over 1 day for two stand alone systems: lighting kit (above) and hybrid system with photovoltaic panels coupled with a diesel generator (below)*

Autonomous photovoltaic systems have particularly penalizing restrictions that are essentially linked to the absence of charge control (being directly dependent on the amount of sunshine) but also linked to daily cycles (generally charging in the day and discharging at night).

3.2.3. *The market for photovoltaic systems*

The market for photovoltaic systems is divided into several segments according to use, size, and locality. For these markets, different technologies or battery designs can be used, allowing specific differences in terms of size and resistance to functioning constraints.

Generally, we can distinguish professional applications from domestic applications.

Domestic applications are concentrated essentially on rural electrification where there are lighting and telecommunications needs. While these applications have been rapidly saturated in developed countries (with systems that are isolated from the electrical network having been rapidly provided for), developing countries constitute the most important part of the market. Thus, world energy analyses from 2004 showed that a quarter of the world population (2.6 billion individuals) do not have access to electricity. These populations are mainly located in provinces in South Asia and in sub-Saharan Africa. The EPIA (*European PV Industry Association*) anticipates a capacity of 30 GWp of photovoltaic energy being installed between now and 2020. Installation campaigns are generally launched and funded by diverse organizations (such as the World Bank).

Professional systems correspond to the needs such as telecommunication, pumping, cooling, and finally supplying a variety of weakly powered devices (signaling or street lights). By contrast to the preceding market, professional systems present regular growth, even accelerated for street lighting, without the benefiting from funding.

For these two applications, the cost of the storage system represents a quarter of the cost of the entire system (for example, €3/Wc out of a total system cost of around €12/Wc). However, as shown in Figure 3.3, the costs associated with batteries over 20 years can account for half the costs of the system due to the short lifespan of batteries.

As a result, storage is a crucial factor for autonomous photovoltaic systems. Two types of market are, therefore, to be considered: the installation market and the replacement market.

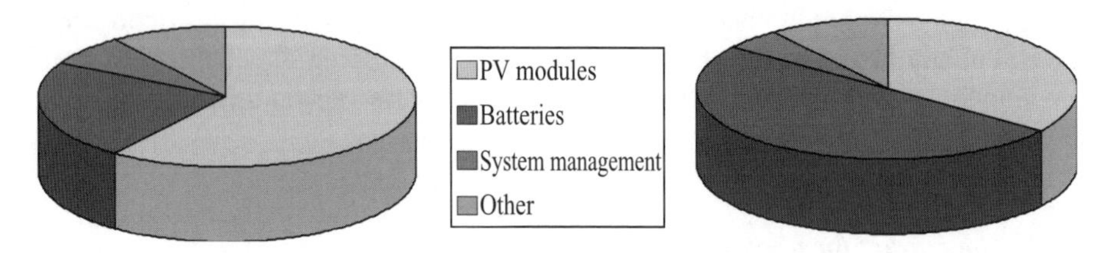

Figure 3.3. *Photovoltaic system costs broken down. Left: investment costs; Right: possession costs over 20 years*

On the world scale, all systems must reach 130 GWc by 2030 with the following division:

– 70 GWc for professional systems;

– 60 GWc for rural electrification.

3.2.4. *Sizing storage for autonomous photovoltaic systems*

Sizing the storage for a photovoltaic system is essentially done based on the energy needs of the application. This sizing can be split up into several steps:

– First of all, an initial analysis will enable the system consumption to be optimized by, for example, changing certain electronic devices (such as lower energy light bulbs).

– Then, an evaluation of energy needs is carried out. This evaluation is easier for professional applications, for which energy needs are more easily listed than for domestic applications where individual behavior presents a level of randomness. Moreover, in all domestic applications, these needs have a tendency to increase rapidly when additional devices are added, and this encourages the use of modular autonomous photovoltaic systems.

– The third step in sizing the system concerns the choice and sizing of the generator. Sometimes, hybridization of solar panels is considered (with an electrical generator for example), especially when there are very marked seasonal variations (alternation between dry and rainy seasons). The size of the photovoltaic generator will then be calculated, taking into account the local solar resource and the efficiency of the chosen module. Energy production should be equivalent to the daily energy consumption.

– Finally, and only after all these steps, there is the sizing of the storage. This corresponds to multiplying the number of days of autonomy (or the number of consecutive days with little sunshine) by the daily energy required. This quantity of energy will be limited by the maximum depth of discharge acceptable for the selected technology.

3.2.5. *Choice of appropriate storage technology*

The choice of this technology for autonomous photovoltaic applications is a compromise between different factors:

– cost is often a primary factor and describes either the cost of investing in the technology, or the global cost comprising the lifespan and maintenance;

– energy efficiency: this is an essential factor especially when production costs are elevated. If energy efficiency of the storage technology is less than 75%, this will require over-sizing the PV modules by 25%;

– charge retention: this factor is a combination of the efficiency and the self-discharge characteristic of the technology. It enables determination of the amount of energy remaining in the battery after a given period;

– maintenance: this factor will significantly impact the global cost of the system, especially for very remote systems;

– adaptability to variable operating conditions: the lifespan of batteries can be affected by temperature and the type of cycles undertaken (deep or partial cycles, discharge or charge current regimes);

– security;

– recyclability.

Different applications of autonomous systems call for different battery technologies in order to guarantee the best possible service.

Among these technologies, the lead acid battery, although in use for more than 100 years, currently offers the best answer in terms of price and availability. Some sites, where the climate is particularly severe, presenting extreme working temperatures, may be equipped with nickel-cadmium batteries, but their elevated purchase cost (around €400/kWh) prevents their more general employment.

For peak power installations, close to 100 W, lead acid batteries with flat plate electrodes are generally used (Figure 3.4). These batteries are not very expensive to purchase (€50-60/kWh), but are not very reliable (lifespan of 6 months to 4 years) suggesting variable overall costs (from €0.4/kWh to €0.8/kWh). More significant installations use batteries with tubular electrodes (stationary technology), which are better adapted to daily cycling but have a greater purchase price (€100-250/kWh; Figure 3.5). This type of battery is used in installations of several hundred watts to several kilowatts at peak time, and in all professional applications due to its reliability and security (television and radio relay stations, telecommunication relay stations, lighthouses).

The lifespan of these technologies is much higher than the preceding technology (from 4 to 12 years). The overall storage cost remains around €0.50/kWh.

The watertight lead battery using recombination of gases is used in restricted environments where maintenance is rare, such as maritime beacons, or confined installations (Figure 3.6). It is the most expensive lead technology (€150-300/kWh) for limited lifespan (around 6 years) leading to quite high overall costs (around €1/kWh).

Figure 3.4. *Lead acid batteries for car applications*

Figure 3.5. *Stationary lead acid batteries*

Figure 3.6. *Spiralled lead acid batteries*

3.3. Limited lifespan for lead acid battery technology

Being particularly sensitive to operating conditions, the lifespan of lead acid batteries are generally shorter for outdoor operations than for certain dedicated applications (stationary, or car applications) but are especially difficult to predict.

Several specific degradations of photovoltaic applications can be observed linked to restrictions in their function [MAT 03]:

– stratification of the electrolyte (much higher acid concentration in a small part of the battery) favored by the daily cycles of the battery;

– hard sulfation of positive and negative electrodes due to the battery's low state of charge and also due to stratification of the electrolyte (in that case lead sulfate is located at the base of the electrodes, see Figure 3.7);

– corrosion of the cathode encouraged by overcharging and due to the battery operating at high temperatures (Figure 3.8);

– Shedding of the active mass, especially for technologies with flat plate electrodes (Figure 3.9).

Figure 3.7. *Micrography of irreversible lead sulfate crystals on the positive electrode of a lead battery*

Figure 3.8. *Micrography of a corrosion layer on the positive electrode of a lead battery*

Figure 3.9. *Picture of a positive lead acid battery electrode showing shedding of active material*

3.3.1. *Battery energy management*

Insofar as recharging of stand alone photovoltaic systems occurs as a function of the amount of sunshine present, energy management becomes a crucial point, at the same time guaranteeing the best service for the user, while staying within the lifespan of the system.

Such energy management is carried out by the regulator, which will enable the battery to function at states of charge adapted to the technology. The majority of regulators currently used commercially were developed for lead acid technology and very rarely for nickel-cadmium technology. They are generally based on a measure of the voltage and current of the battery.

The most common principles use four different voltage thresholds (Figure 3.10):

– HVD (*High Voltage Disconnect*) to stop charging;

– LVD (*Low Voltage Disconnect*) to stop discharging;

– two intermediate thresholds that enable reconnection of the battery after relaxation: HVR (*High Voltage Reconnect*) and LVR (*Low Voltage Reconnect*). These thresholds are determined taking into account the sizing of the system.

Energy management strategies used fall into two categories corresponding to either a daily management, or a periodic management allowing eventual rehabilitation of certain degradations.

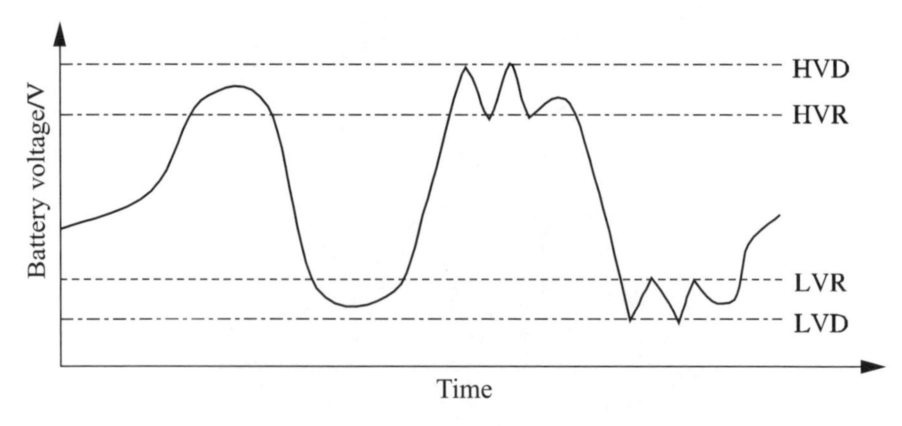

Figure 3.10. *"All or nothing" principle of function and of regulation of batteries used in autonomous photovoltaic systems (HVD, HVR, LVR, LVD)*

The different strategies for daily management are as follows:

– "All or nothing" management of charge and discharge: this very rudimentary strategy consists of disconnecting the battery when it reaches different thresholds, with reconnection only occurring once the battery has relaxed and reconnection thresholds have been reached. Figure 3.11 illustrates the impact of the value of the HVD threshold on the daily energy delivered by the battery, showing the important influence of the values chosen for different thresholds.

– "Floating" charge management: in this case a weak charge current is maintained after reaching the disconnection threshold (HVD) and the current is generally weaker than in the previous mode of management. This current is then adapted to keep the voltage of the battery constant. There is no need for a reconnection threshold (HVR).

– "Pulse modulation" management: some regulators use new electronic components that enable modulation of the recharge current and adapt its period of application, i.e. they cut off.

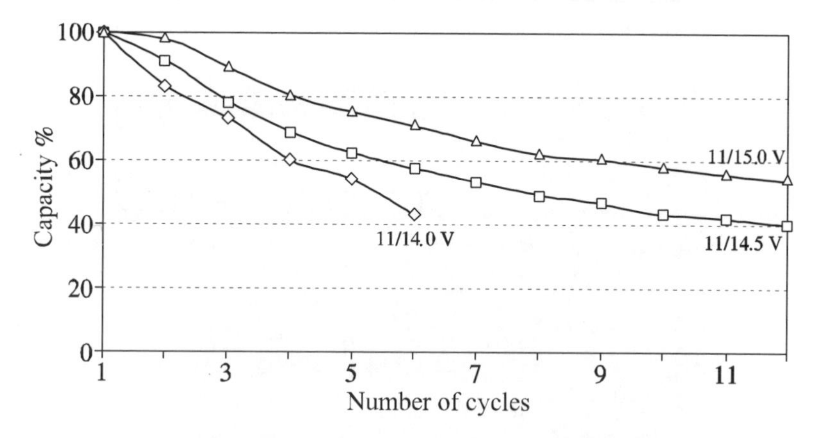

Figure 3.11. *Influence of HVD and LVD thresholds on the energy delivered by a lead acid battery*

The benefits of modulated or pulsed recharge current was demonstrated several years ago [SRI 03, KIR 07], limiting the parasitic effects of water electrolysis and improving the efficiency of recharge. The main issues concern signal optimization (frequency and cyclic ratio of the pulses) while limiting the electrochemical degradation of the battery. Laboratory results should lead to marketing of optimized regulators within 2 years from now.

Periodic management of autonomous photovoltaic systems consists of improved recharging by temporarily increasing the HVD threshold from 14.4 V to 14.8 V for an open lead battery of 12 V. This protocol, called "boost charge", was adapted specifically for lead acid technology, and aims to improve the recharge and diminish the stratification phenomenon. It avoids the frequent overcharging that can cause loss of water and corrosion of the grid of positive electrodes.

The periods of application of such "boost charges" varies from one regulator to another: it can only be applied initially, periodically (every week or every 10 days, for example) or else set off manually. The disadvantage of this method is once again in the absence of charge control for the photovoltaic application: when the protocol of "boost charge" is launched, the energy necessary to complete the protocol is not always available.

Finally, a predominant factor is the development of the charge and health indicator. The regulators can be more and more sophisticated and include information on the energy available instantly (SOC for "state of charge") or on the total energy available (SOH for "state of health") [DEL 06]. This information is very useful for users, and can also help to improve the overall intelligence of photovoltaic systems and increase battery life (Figure 3.12).

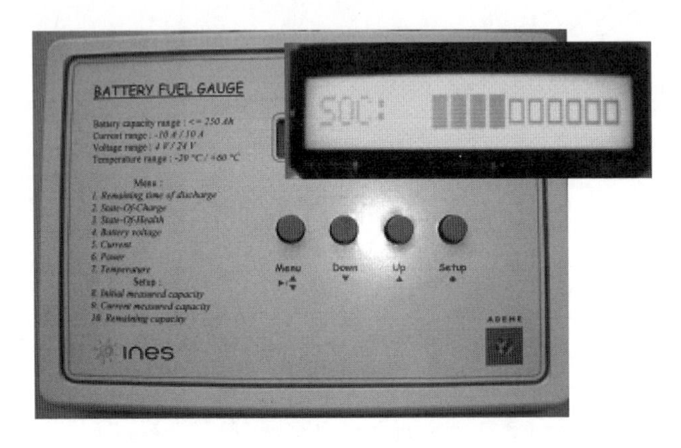

Figure 3.12. *Control panel of a lead battery equipped with a SOC gage [DEL 06]*

3.3.2. *Lithium-ion technology seems promising*

What are the future issues for autonomous photovoltaic systems storage?

Figure 3.13, representing trials with different technologies (lead, nickel-cadmium and lithium-ion) shows the interest in lithium-ion technology for this application, taking into account its cycling characteristics. This technology presents further advantages: high energy efficiency, elevated lifespan, absence of maintenance, reliability, predictable behavior. The cost of accumulators, which is the main limiting factor, is currently reducing (gain of a factor of four over the last few years, a tendency confirmed by noting their use in hybrid and electric vehicles), so this technology should play an important role in the next few years.

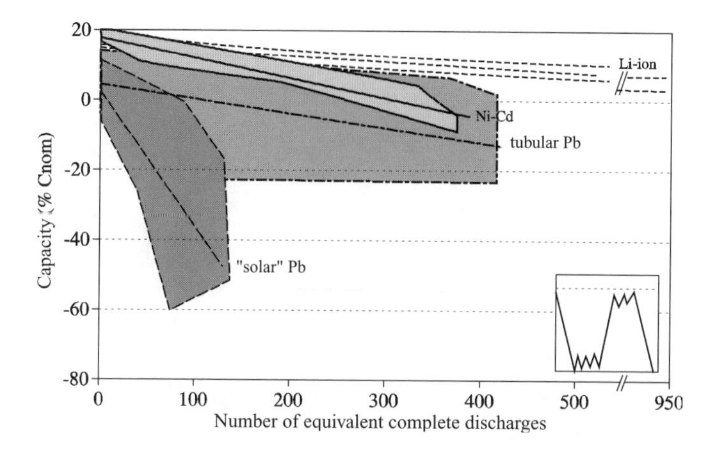

Figure 3.13. *Losses of capacity when cycling (in %) for different battery technologies (lead, nickel-cadmium, lithium-ion) according to the applicable norm in accumulators used in autonomous photovoltaic systems [LEM 08]*

For the past 2 or 3 years, several research projects aimed to introduce this technology to photovoltaic systems [PER 06, MAT 07] by including storage modules of several tens of Ampere-hours, while adapting the sizes of these systems and their management.

First of all, professional applications (maritime beacons, street lighting) for which lithium-ion is already competitive are considered with predominant reliability criteria and usage costs of €0.2/kWh (compared to €0.5-2/kWh for lead technology).

Lithium-ion technology will probably be the solution to attain storage lifespans equal to those of the photovoltaic modules, namely 20 to 25 years.

3.4. Grid connected systems

3.4.1. *An evolving electric network*

For the past 5 years, upheaval of the electric networks of developed countries has occurred due to several factors:

– the will to limit energy sources that emit CO_2, encouraged by European directives (white book) or world directives (Kyoto protocol);

– the liberalization of the electricity market, inducing uncertain returns on investment for traditional means of production, and therefore, promoting a multiplicity of decentralized production sources (Figure 3.14).

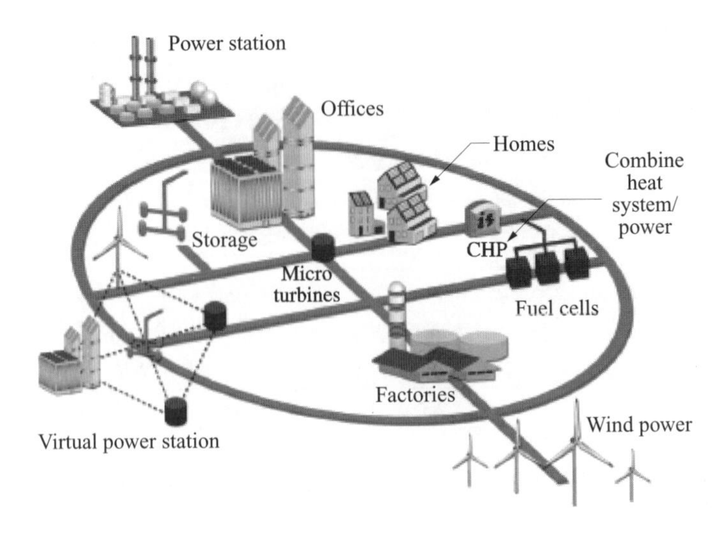

Figure 3.14. *Configuration of the electric network showing decentralized production sources*

In the context of decentralized distribution, electricity storage has a vital role to play. Beyond its utility in compensating for variations in electricity production, it enables adaption to the demand by at any moment injecting previously stored energy into the network. It is a temporal vector of electricity.

In addition, we distinguish different aims of storage for different needs or different periods of storage (Table 3.1). These aims cover all potential applications of storage systems.

Applications	Discharge duration	
	Minimum	Maximum
– Peaks shaving	4 hours	10 hours
– Transmission support	2 seconds	5 seconds
– Demand management	4 hours	12 hours
– Current quality	10 seconds	1 minute
– Security	15 minutes	5 hours

Table 3.1. *Example of applications for storage connected to the network and associated discharge durations*

3.4.2. *A multiplicity of storage systems for different functions*

Different storage technologies are required to provide the power, response times and discharge durations required by systems connected to the network.

For low power, of the order of 10 kW at peak, batteries, especially lithium-ion technology, are most suited for security or peak-shaving applications (Figure 3.15). The supercapacitors will be the same as those used with profiles of microcycling for applications of current quality.

Figure 3.15. *Example of a lithium-ion storage system for a solar habitat connected to the network*

For greater power, of the order of 100 kW at peak times (as for a factory or a housing estate, for peak shaving), sodium-sulfur batteries (Figure 3.16) functioning at high temperature or redox flow systems (Figure 3.17) are of interest due to their cyclability [VRB, AIE].

Figure 3.16. *Sodium-sulfur batteries*

Figure 3.17. *Electrolyte redox batteries*

Finally, for very high power, of the order of 1 MW at peak times (for a power station), storage systems initially require heavy investments for constructing an infrastructure (pumping for hydraulics, thermal storage, air compressors coupled with gas turbines [MUL 07]), which are economical because of the low additional costs (Figure 3.18).

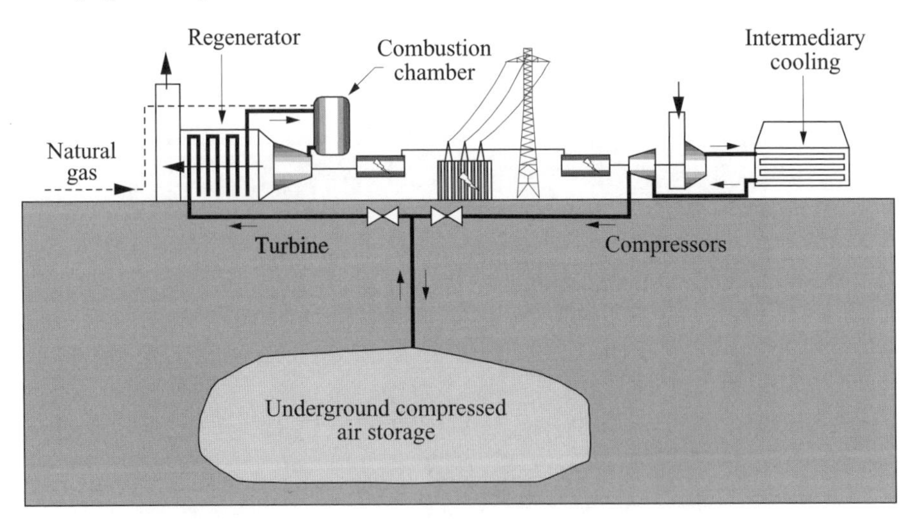

Figure 3.18. *Cavern equipped with air compressors coupled with gas turbines*

3.4.3. *Storage linked to the network; a big issue for the energy sector*

The increasing popularity of photovoltaic systems connected to the network requires the use of all kinds of storage technologies: conventional or advanced batteries, mechanical storage systems of different powers. The fit between technology and application is a primary area of research. An equally predominant area is the development of adapted bidirectional electronic interfaces (allowing reinjection of the battery on the electric network, or a recharge of the battery from the network), which present adapted voltage levels and also counters (to count tariffs).

Finally, depending on where storage systems are located on the electric network (generation, transmission, or distribution), an approach that identifies the benefits connected with these systems and for each of the agents (producer, transmitter, and distributer) is indispensible as it will enable the final selection of different energy solutions.

3.5. Bibliography

[AIE] www.aie.org.au/syd/downloads/vassallo.pdf.

[DEL 06] DELAILLE A., PERRIN M., HUET F., HERNOUT L., *Journal of Power Sources*, vol. 158, no. 2, pp. 1019-1028, 2006.

[KIR 07] KIRCHEV A., DELAILLE A., PERRIN M., LEMAIRE E., MATTERA F., *Journal of Power Sources*, vol. 170, pp. 495-512, 2007.

[LEM 08] LEMAIRE-POTTEAU E., MATTERA F., DELAILLE A., MALBRANCHE P., "Assessment of storage ageing in different types of PV systems: technical and economical aspects", *23rd European PV Solar Energy Conference*, Valence, September 2008.

[MAT 03] MATTERA F., BENCHETRITE D., DESMETTRE D., MARTIN J.L., POTTEAU E., *Journal of Power Sources*, vol. 116, pp. 248-256, 2003.

[MAT 07] MATTERA F., MERTEN J., MOURZAGH D., SARRE G., MARCEL J.C., "Lithium batteries in stand alone PV applications", *22nd European PV Solar Energy Conference*, Milan, Italy, September 3-7, 2007.

[MUL 07] MULTON B., RUER J., *Stockage de l'énergie électrique: état de l'art,* Diaporama ECRIN, February 15, 2007.

[PER 06] PERRIN M., MALBRANCHE P., LEMAIRE-POTTEAU E., WILLER B., SORIA M.L., JOSSEN A., DAHLEN M., RUDDELL A., CYPHELLY I., SEMRAU G. *et al.*, "Comparison for nine storage technologies: results from the INVESTIRE Network", *Journal of Power Sources*, vol. 154, no. 2, pp. 545-549, 2006.

[SRI 03] SRINIVASAN V., WANG G.Q., WANG C.Y., *Journal of Electrochemical Society*, vol. 150, pp. A316-A325, 2003.

[VRB] PRUDENT ENERGY, Prudent Energy Inc., online at: www.vrbpower.com, 2007-2009.

Chapter 4

Mobile Applications and Micro-Power Sources

This chapter aims to be an introduction to the highly dynamic world of energy sources for onboard applications. Given the rapid progress of research in this emerging domain, we have decided to present a representative overview of the very diverse current state of the art. Although not exhaustive, this overview is wide and, therefore, does not go into the details of each case that is presented. Instead, we direct the reader to numerous references with which he will be able to expand his understanding of each area.

4.1. The diverse energy needs of mobile applications

Mobile or onboard applications have additional restrictions compared to traditional applications. Due to its mobility, the system being supplied will be disconnected from the electric sources that are usually available to it. Therefore, it is necessary to place the energy source onboard, and this has a cost, in volume as well as in weight: it is this constraint above all others that leads to the need for μ-sources.

Several independent criteria can be used to classify μ-sources:

– level of energy;

– level of power;

– primary source (generation) versus secondary source (conversion);

Chapter written by Jérôme DELAMARE and Orphée CUGAT.

– storage versus recovery of ambient energy;

– remote supply, etc.

We can choose to distinguish two families according to the "scale" of their power needs.

4.1.1. *"Weak" powers (su-Watt wattage)*

The consumption of very weak power systems starts at the microwatt scale. This is the case, for example, for scattered or onboard sensors, with the goal of capturing and transmitting information: to monitor a parameter and transmit information from time to time to a remote center. As a result of micro-technology – and now of nanotechnology – the performance of sensors has evolved well in terms of consumption. The consumption of commercial sensors now is in the neighborhood range from ten to 100 mW.

Microsensors are being developed that are not only of reduced size (the complete size of the chip is going from a square millimeter to a tenth of a square millimeter) but also with consumptions that have fallen drastically. Detections that are based on deformation today allow the design of sensors (accelerometers, gyrometers, field sensors, etc.) whose consumption is continuous and in the order of microwatts. Progress has also been made in electronics for processing and transmitting information. Currently important questions are, for example, should information be processed locally and then transmitted discontinuously, or should raw sensor information be transmitted continuously and processed far away?

All specifications are different and applications that are simple *a priori* can involve energy problems. For example, let us take the case of the autonomous push button which "only" has to transmit "stop/go" information to the equipment being controlled. This button will draw its energy from its environment, by converting the energy produced by the person pushing the button into electrical energy. To start with, the user will only wish for "stop/go" information; then, he will wish to be able to code this push-button in order to be able to distinguish it from other interrupters; finally he will want the push-button to be able to receive confirmation that the system has received the information and, when necessary, to retransmit the information. The second case requires a larger source of energy than the first case, since the last case requires not only a larger source of energy, but also a storage device in order to be able to operate differently when needed.

Evidently, applications are diverse, and are not limited to sensors. A biomedical instrument, such as an implanted insulin micro-pump or a pacemaker are other examples. The diagram below (Figure 4.1) gives the orders of magnitude of the power necessary for different systems to function.

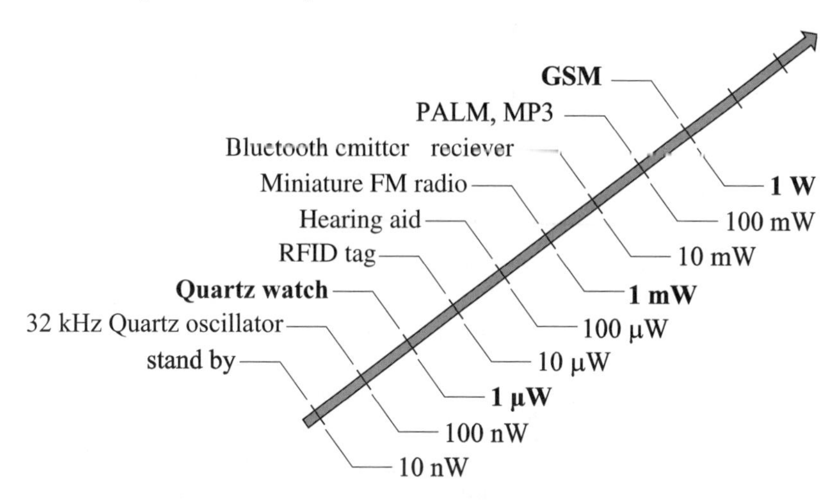

Figure 4.1. *Range of powers required by different electronic equipment*

The onboard system may equally require a large amount of power (a few Watts) but consumed as pulses or in a temporary way: a mobile phone consumes 1 to 3 W when being used but only several milliwatts when on standby. For autonomous applications that function in a discontinuous way, the power will not always be the main criteria: the fundamental factor will then be the energy necessary for function (see section 4.1.3).

4.1.2. *"Large" powers (a few Watts)*

Greedier applications frequently require power in the order of 10 W: onboard micro-robotics, micro-drones (Figure 4.3), high-tech equipment (portable computer, etc.). In micro-electronics the performance of components evolves according to Moore's Law, and as the number of functionalities in a laptop multiplies the consumption is increased, but unfortunately Moore's Law does not apply to batteries (Figure 4.2)! Even if alternatives to batteries are found, with systems such as micro-fuel cells, the gains are only one order of magnitude greater than the current levels. So, for mobile sources following Moore's Law is problematic (or "more than Moore"). The increased performance of components should, therefore, take place without the need for supplementary consumption.

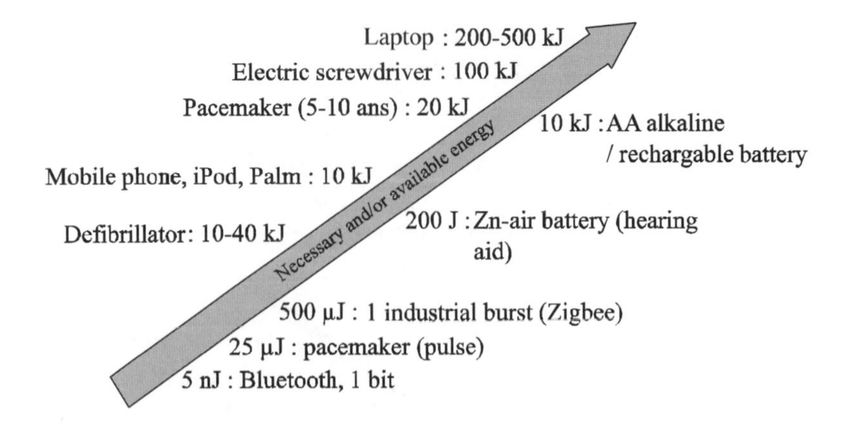

Figure 4.2. *Energy necessary for certain mobile applications and energy available from certain "established" sources*

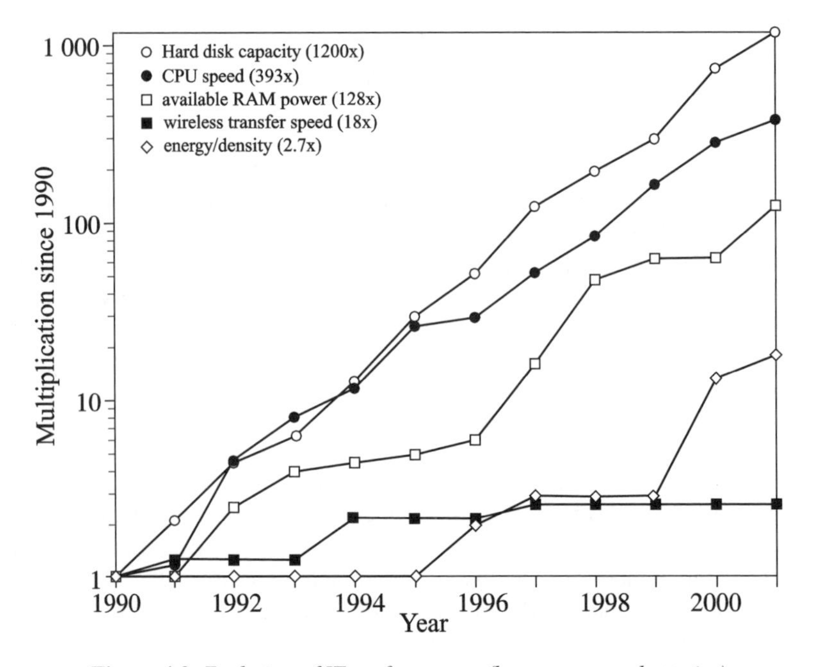

Figure 4.3. *Evolution of IT performances (lowest curve = batteries)*

4.1.3. *Energy needs*

In the previous two sections, the needs were described in terms of power. However, for mobile systems, energy is a parameter that is often more predominant (!) than power. It is the parameter that defines the range of the system. The energy needs of mobile systems stretch over several orders of magnitude going from the nanojoule (nJ) to the megajoule (MJ). Moreover, the units commonly used are different. In the case of wireless transmission, nanojoules are used, whereas for batteries of a laptop computer, the energy is expressed in ampere-hours or milli-ampere-hours linked to a voltage.

At the low end of the consumption scale, we will, therefore, find the energy necessary to transmit information. In the case of wireless transmission, the consumption today is around 5 to 20 nJ per bit transmitted. In the case of an industrial device, which must transmit several octets of information, coded with redundance, across several tens of meters, the energy needed is in the neighborhood of 500 μJ. These orders of magnitude correspond to industrial wireless transmissions in systems, such as ZigBee.

Very weak consumptions may nevertheless require significant energy storage. A cardiac stimulator, for example, will consume 25 μJ per pulse. However, with a lifespan of greater than 10 years, the energy stored in the battery is around 20 kJ (2.8 V, 2 Ah).

In the case of a UAV (unmanned aerial vehicle) or a laptop computer, for example, there is a compromise between power and storage, according to the range required. A computer generally requires 2 to 6 hours of range depending on the use. The corresponding energy is several hundreds of kilojoules (180 kJ at 4.7 Ah, 11.1 V for a Dell D600). A UAV will be able to remain in flight for 10 minutes to 2 hours depending on the mission and the terrain, but with a weight restriction, which is much more draconian than for a laptop. In this case the important criterion is the specific energy.

As an example, for lithium polymer batteries the specific energies that are currently attainable under "severe" use are around 140 Wh/kg. In cases where this specific energy is not sufficient, it is possible to achieve a gain on condition that recharging may not be possible: using zinc-air cells, the specific energy can reach 370 Wh/kg (these cells are currently used in auditory equipment).

4.1.4. *Adequacy for the duration of the mission*

As we have seen, the choice of the type of source from amongst the diverse possibilities available depends on numerous criteria for the appropriateness of the source for the type of use. However, another type of choice, which depends on the predicted duration of use, is worth examining.

Figure 4.4. *UAV CPX4 (40 W, 60 cm) [NOV]*

Let us take the example of the electric motorization of a miniature drone, which requires 15 W of continuous power. Depending on whether it must function for 10 minutes, 30 minutes, or 1 hour, the choice of the source may differ. In the case of a battery, for a given type of technology, power and stored energy are directly proportional to the mass or to the volume. For a source that uses fuel, the required power determines the mass and the bulk of the converter, but it is the duration of the mission that determines the quantity of fuel that should be put onboard. Depending on the efficiency of the source and of the converter, and the type of fuel onboard, filling the capacity will be more or less heavy and bulky.

Figure 4.5 illustrates this reasoning: in order to simply ensure the maintenance of flight, only the combined mass of the generator and the converter, which are capable of supplying 15 W, needs to be considered. Then, the quantity of fuel for the chosen duration must be accounted for, depending on the efficiency of the system. We note that in this special case, it would be more worthwhile to use a lithium-ion battery for very short missions (a few minutes), and a fuelled micro-turbine coupled with an electric generator for missions between 8 and 20 minutes, and finally beyond 20 minutes, it is necessary to choose a thermo-electric micro-generator, which is heated by fuel. Here we are undertaking a simplified theoretical study based on predictions of system performance. This example does not aim to obtain a categorization of the different sources that have been cited, but to list the various parameters that must be taken into account when matching the source to the mission.

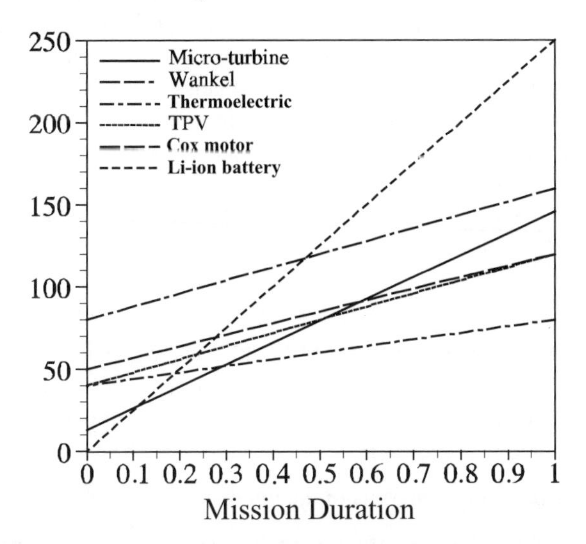

Figure 4.5. *Comparison of the onboard mass (grams) of various viable sources, for a power of 15 W, as a function of the duration of the mission*

These figures are only illustrative: the data used for calculations (powers, efficiencies, size) is drawn from a non-exhaustive analysis of a selection of publications in 2003 and, therefore, has evolved since then. The preferred solutions are probably very different today!

4.2. Characteristics due to the miniaturized scale

This work is about "storage": primary sources "conform" the best to this concept. Therefore, we will rediscover the majority of the solutions that are traditionally given at the macroscopic scale. Energy may be stored in a form that is directly usable: electric (super-capacities) or electrochemical (cells or batteries). It can also be stored in a form that must be converted into electricity in a separate stage: here we typically mean chemical energy (hydrocarbons, hydrogen, pyrotechnics, etc.), and sometimes radioactive energy. This is the case for μ-converters rather than for μ-sources.

Due to the small size of onboard systems, some solutions are not viable (for example, hydroelectric μ-dams are not yet being studied!), but new solutions do become foreseeable.

For some applications that are not greedy, we can even envisage recuperating the wasted energy present in the environment and converting it, *in situ*, into electricity. We can conceive a μ-converter adapted to the type of "free" energy that is potentially available: light (sunshine, lighting), vibrations, and mechanical deformations (machines, motors, structures, humans), sources of heat (sun, humans, motors), pneumatics (variations in atmospheric pressure, wind speed, compressed air leaks, the front edge of an airplane wing)…the list is long!

We note that, we can exploit both thermal gradients and/or the temporal variations of a homogenous temperature.

Miniaturization of all these solutions poses several major problems:

– laws of scale reduction evolve differently according to the principles being brought into play. So, the thermodynamics of reactions and heat transfers over such small sizes pose problems in the design of structures and in the choice of resistance for materials;

– integrations of exotic materials in the micro-technology processes;

– micro-fluids and micro-tribology at high speed and small dimensions;

– the efficiency of all these systems is very dependent on the fabrication technology;

– and, of course, even when a realistic fabrication method is found, there is the price…

Fitting such a quantity of information in one chapter is impossible. We therefore encourage the reader to consult the literature, in particular some very well-written reviews [COO 08; JAC 02; ROU 04; YEA 07; KAR 08; MAT 08].

4.3. Capacitative storage

Supercapacitors present better performances than batteries in terms of instantaneous power (as much in charging as when discharging), and better performances than capacitors in terms of specific energy. Therefore, they represent a compromise combining promising assets for "mobile" storage.

Numerous works are underway regarding their integration and these principally concern the development of specific materials for electrolytes and electrodes [LIU 08; KIM 02; KAM 07].

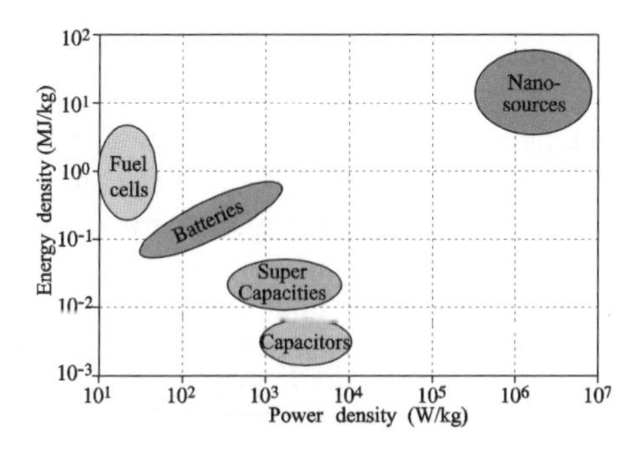

Figure 4.6. *Ragone diagram comparing various storage systems (electric) as well as nano-energetic materials (thermal, pyrotechnics)*

4.4. Electrochemical storage

We can list three families: cells, batteries (or accumulators), and fuel cells. The literature is full of information on their miniaturization, and we will content ourselves with mentioning their integration on silicon, which is being worked on by numerous laboratories. Figure 4.6 compares various electric storage solutions.

4.4.1. *Cells*

When recharging is not required, cells offer a good solution in terms of storage. Alkaline cells, greatly used by the public, offer specific storage performances similar to the best nickel-metal hydride (NiMH) batteries. As for zinc-air batteries, they enable us to obtain specific energy densities which are almost three times more than for lithium-polymer batteries. The table below gives the orders of magnitude of the specific energies that can be obtained using different miniature cell technologies.

Ref.	Type	Dimensions (mm)	Voltage (V)	Charge (mAh)	Energy (J)	Weight (g)	Energy density (Wh/kg)
LR60	Alkaline	Ø6.8 × 2.15	1.5	15	81	0.27	83
SR416	Silver oxide	Ø4.8 × 1.65	1.55	8	45	0.13	95
HA10	Zinc-air	Ø 5.8 × 3.5	1.4	90	450	0.31	400
(Military)	Lithium thionyl chloride	Ø 14.2 × 50	3.6	2,400	31,000	(25?)	340

Table 4.1. *Specific energy provided by various miniature cells*

4.4.2. *Batteries and accumulators*

The technologies and performances of batteries are described in other chapters in this book. However, for mobile applications, the use of high-performing technologies is more easily justified by economic criteria.

An iPhone of 130 g sold at €800 (without the subscription) at the start of 2009, comes to €6,000/kg. A Twingo of 850 kg sold for €7,000 at the same time comes to €8/kg. In the first case, the use of a lithium polymer battery is the best choice, whereas in the second case a lead battery should be chosen in order to be able to start the motor.

An order of magnitude for the performances of these batteries is given in the summary table below.

Type of battery	Energy density	Power density	Number of cycles
Lead (Pb)	20 to 40 Wh/kg	Up to 1 kW/kg	500 to 1,200
Nickel-cadmium (NiCd)	50 Wh/kg		2,000
Nickel-metal hydride (NiMH)	75 Wh/kg		500 to 1,000
Lithium	100 to 150 Wh/kg	Up to 4 kw/kg (30°C)	Up to 1,200 (often much more)

Table 4.2. *Power density, specific energy, and charging cycles for various types of batteries*

In small dimensions, there exist technologies that are different to those that are usually used.

Miniaturization of these batteries and their integration on silicon is a very active area of research [DAN 02; SAL 08; NAG 04; EFT 04]. The goal is to be able to directly link the battery to its electronic circuit on the same substrate.

For this goal, two main paths have been explored. The first consists of setting up the battery on a supple substrate and then placing that on the electronics. The second involves depositing the layers that will make up the battery directly on the silicon.

Figure 4.7. *Batteries that are integrated on silicon (CEA)*

The capacitances for lithium batteries placed on silicon go from 100 to 400 μAh/cm^2, for voltages of 3.8 V. These batteries can provide a maximum current of 1 to 5 mA/cm^2 and can be cycled more than 10,000 times (data from CEA).

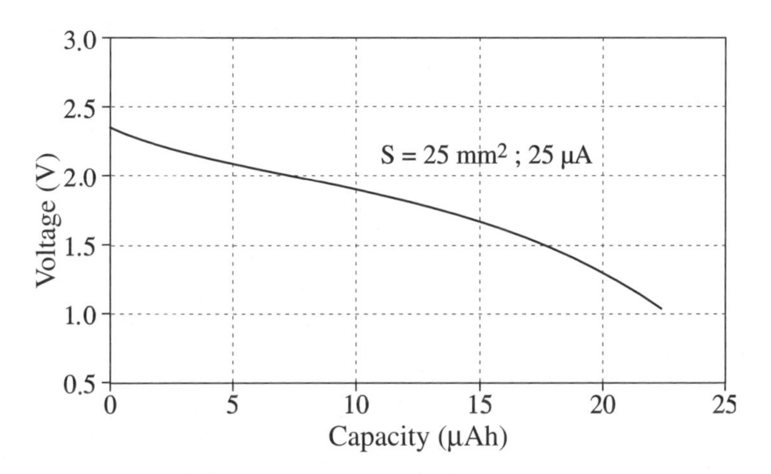

Figure 4.8. *Discharge curve for a micro-battery integrated on silicon*

Apart from their planar layer technology, these batteries currently have exchange surfaces that are equal to the surface occupied on the substrate. In order to improve their characteristics, the following battery developments exploit the third dimension:

by placing layers in "micro-concertina" form, researchers in this field hope to increase the exchange surfaces by 10 times and similarly to improve performances by that factor [BAG 08].

While working on compressed nano-powders and composite electrodes, the improvements in ionic exchanges have enabled researchers at MIT (Massachusetts Institute of Technology) [KAN 09] to obtain discharge times of around 10 to 20 seconds (from 200° C to 400° C). Such performances allow the energy stored in lithium polymer batteries to be combined with the discharge rates of supercapacitances.

4.4.3. *Fuel cells*

Fuel cells constitute an attractive solution for certain ranges of onboard energy: the diverse solutions developed are described and compared in recent reviews [KUN 07; MOR 07].

The Fraunhofer Institute for reliability and micro-integration (IZM) and the Technical University of Berlin (TU) have collaborated in developing a fuel cell that provides 12 W and weighs 30 g. Such a power density (400 W/kg) had never been attained by systems that weigh several hundred grams [BUL].

Other researchers in this field include the CEA (Commissariat à l'Energie Atomique, Grenoble, France) [MAR 05; PIC 07; GON 06], the IEMN (Insitut d'Electronique, Microelectronique et Nanotechnologie, Lille, France) [KAM 08], the University of Sherbrooke [EUR 01], a collaboration between the Universities of Vancouver and Berkeley [CHI 06], etc.

Mobile applications are a potential market for fuel cells. The democratization of current fuel cells for high power (in transport or habitat) is limited by the availability and the price of materials (platinum, nafion). The quantities of materials necessary are smaller for low powers and the costs are comparable with current batteries that are already very expensive.

The French company Paxitech [PAX] manufactures fuel cells that can supply different mobile applications. A fuel cell can generate 20 W and has a mass of 500 g. Such a cell must be linked to a fuel reservoir. A reservoir of metallic hydride, enabling hydrogen to be stored, leads to the generation of 160 to 180 Wh/kg (by the fuel cell). However, laboratory research has shown that by storing hydrogen in chemical hydride reservoirs we can obtain 600 to 800 Wh/kg (from the fuel cell). The performances obtained are therefore up to four times better than the best current batteries.

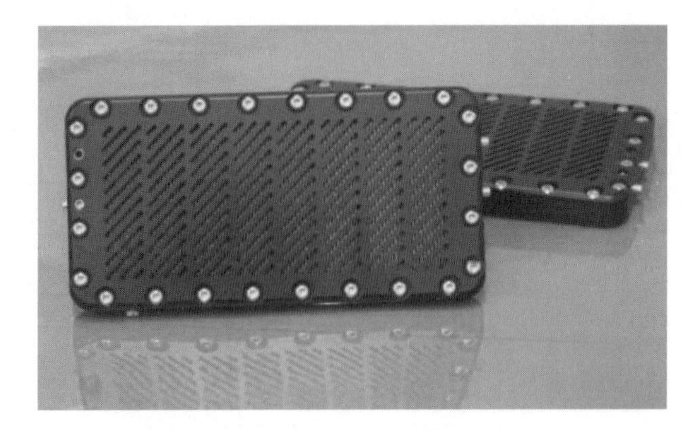

Figure 4.9. *Miniature fuel cell from PaxiTech (conversion module without reservoir)*

4.5. Hydrocarbon storage

The figure below compares the specific energy available in hydrocarbons (before and after combustion) with that available in electrochemical batteries and with mechanics/fluids. It is clear that the energy available in hydrocarbons and in hydrogen is promising, even if we must take into account the low efficiencies in thermodynamic conversion (5 to 20%).

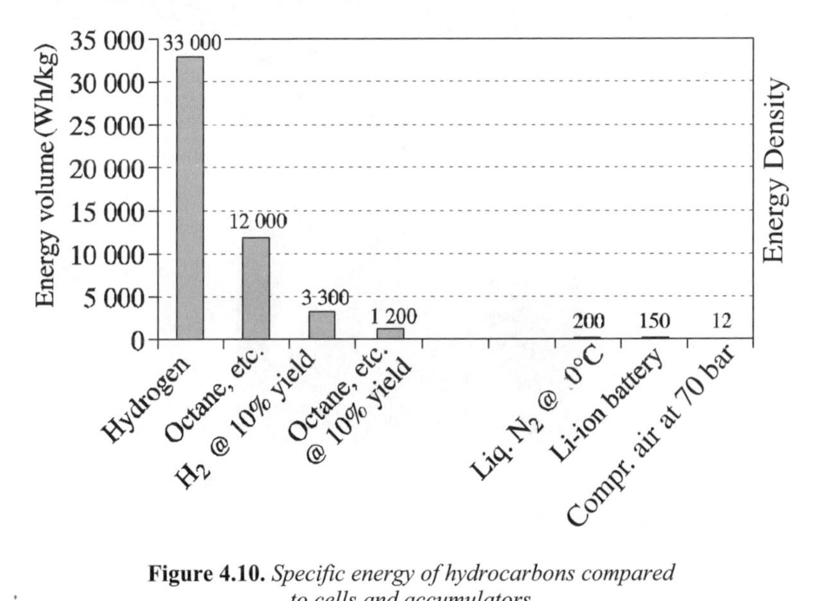

Figure 4.10. *Specific energy of hydrocarbons compared to cells and accumulators*

4.5.1. *Power MEMS*

Power MEMS are micro-systems that generate and convert power. They include systems that convert thermal energy into mechanical energy using combustion of a hydrocarbon in a turbine or a piston motor. The principles being used are directly copied from existing macroscopic applications. Examples are turbo-reactors targeting 3 million turns per minute (MIT), rotating Wankel piston motors (Berkeley), linear free pistons (Birmingham, etc.), reaction exhausts and even steam engines (Sandia), coupled with more or less conventional generators.

4.5.1.1. *Micro-turbines*

The most ambitious project has been developed by MIT over approximately 10 years (*Turbine on a Dime*): a planar turbo-reactor integrated on silicon, which is composed of a pile of almost 10 substrates of silicon. The rotor has a diameter of 8 mm turn and should turn at more than 3 million rotations per minute to produce several tens of mechanical Watts, while ejecting gases at more than 1,500°C.

Figure 4.11. *"Turbine on a Dime": micro-turbine integrated on silicon (MIT)*

Similar studies have been undertaken at the University of Columbia (USA) and the University of Sherbrooke [SHE 02]. The University of Tohoku is also working on turbines that are less ambitious but more realistic. Zwissyg [ZWY 06] and Piers from Leuven Catholic University [PEI 04] have developed less integrated turbines, which attain powers of several tens of watts, based on micro-mechanical technologies rather than micro-technological integration on silicon.

Other projects explore the Wankel rotating piston motor [UNIa; SEN 08] and free pistons [UNIc; UNId].

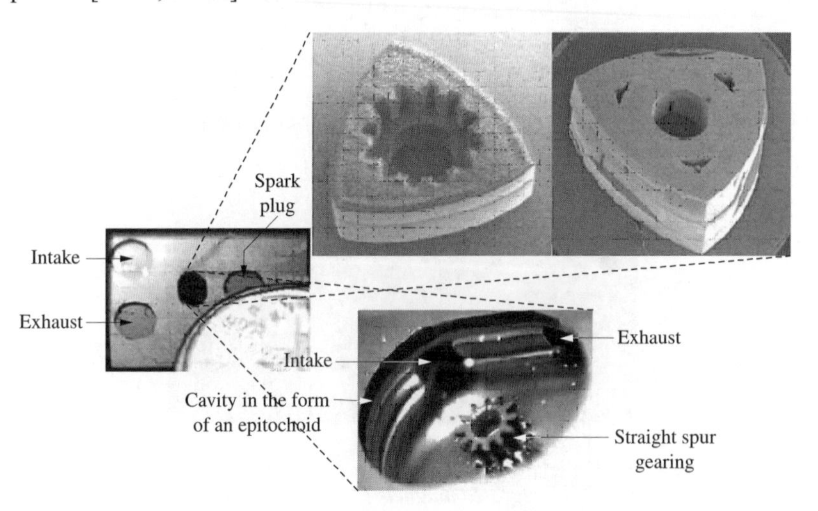

Figure 4.12. *Wankel micro-motor on silicon/SiC*

4.5.1.2. *Mechanic-magnetic conversion*

The generator, which had to be linked to the MIT micro-turbine, was originally an electrostatic induction generator; however, recurring problems of breakdown at 100 V made researchers consider "traditional" electromagnetic conversion, as developed at GeorgiaTech [GEO; HER 08a; HER 08b] using induction.

The result was a series of generators using magnets, which were not specifically dedicated to the MIT "hot turbine".

G2Elab has developed a planar magnetic micro-turbo-generator, mounted on a hybrid magneto-pneumatic bearing, and producing around 50 mW at 30,000 rpm (with the relevant electronic conversion) [RAI 06]. Coupled with a dentist turbine, this generator converts 5 W at 400,000 rpm, and could produce 20 W at 1,000,000 rpm.

Figure 4.13. *Magnetic micro-generators. Top: induction stator (6 mm); bottom: integrated magnets (2 mm; GeorgiaTech)*

Figure 4.14. *Planar magnetic micro-generator with magnets (G2Elab). Photograph by Ch. Morel*

A Chinese team has also been working on this area [PAN 07]. Imperial College London has developed a gaseous flux sensor using the same principle; this sensor is capable of functioning as a weak power generator [HOL 05].

4.5.1.3. *Thermoelectricity*

Thermo-generators use the Seebeck effect (inverse of the Peltier effect). The Seebeck effect is commonly used to measure temperature using thermocouples. A thermocouple is a junction between two metals that produces a voltage when it is subjected to a difference in temperature. The voltages found are of the order of several tens of microvolts per degree Kelvin (50 μV/°K for a thermocouple of type J). By assembling thousands of thermocouples, it is possible to obtain thermoelectric converters generating voltages of the order of 1 V. Commercial thermoelectric converters can reach efficiencies of 6% provided that they have a temperature difference of several hundreds of degrees Celsius between the hot source and the cold source. Unfortunately, conversion efficiency is directly linked to the thermal gradient, and thus decreases drastically as the temperature difference decreases.

Figure 4.15. *LUFO lamp comprising a thermoelectric generator that enables a radio to be supplied*

Assembling thermocouples has been realized for a long time using conventional techniques. New techniques are being developed. Some are based on weaving the metal wires to create thermoelectric fabrics. The conversion efficiencies are still not adequate, but the simplicity can lead to low creation costs. Therefore, it is possible to commercially buy a "petrol" lamp (LUFO) which recovers part of the thermal energy of the flame in order to convert it into electrical energy. The electrical energy produced by this lamp can reach 3 W on some models and can supply a radio.

Using similar principles, studies have been undertaken to recover the energy in a pneumatic for cars with the aim of supplying sensors embedded in the rubber. Warming is produced during driving and creates "hot points". By using thermoelectric converters part of the dissipated energy can be recovered to supply sensors and associated electronics.

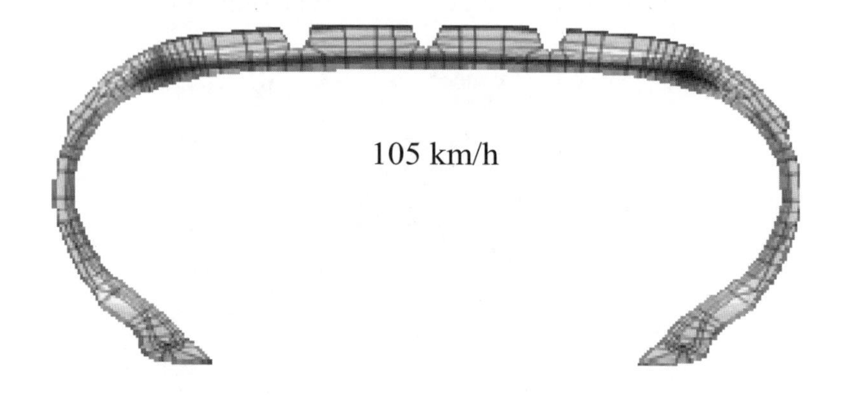

Figure 4.16. *Warming in a pneumatic for cars during driving*

Other types of thermo-generators are also being developed and integrated on networks of micro-thermocouples deposited on silicon [STR 04; SCH 08; WAN 05]. Beyond the material aspect, major difficulties exist in obtaining a significant temperature gradient across small dimensions.

It is extremely difficult to obtain a few hundred degrees or even a few degrees of difference between two junctions that are placed on silicon and spaced at a few hundred microns, especially as silicon is a good conductor of heat.

For relatively high powers, the combustion of propellants or hydrocarbons is used; however, certain systems use body heat to supply watches or "communicating" clothes [LEO 09]. Others are planning systems that can be implanted under the skin [YAN 07].

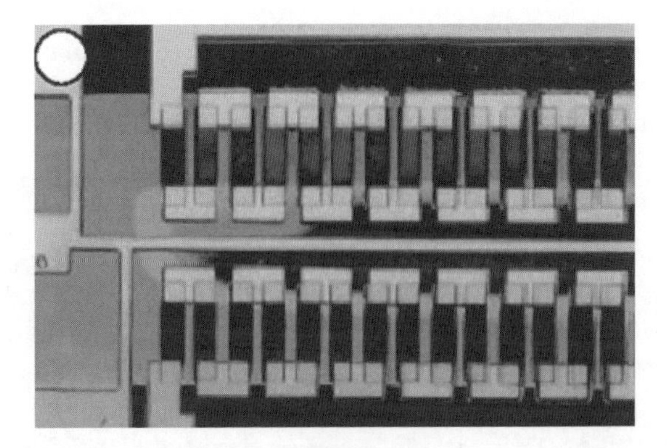

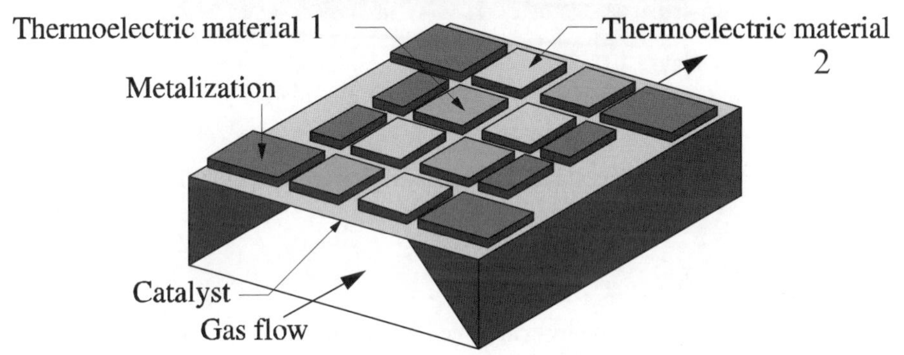

Figures 4.17. *Thermo-electric micro-generator with integrated thermocouples*

4.5.1.4. *Micro-propellers with solid propellant and energy nano-materials*

The principal research institutes in this field are currently LAAS (*Laboratoire d'Analyse et d'Architecture des Systèmes*/Laboratory for Analysis and Architecture of Systems); Toulouse [ROS 05] and the University of Berkeley. These teams are developing pyrotechnical micro-capsules for thermal use (and thus thermo-electric generation) or dynamic use (propulsion) [UNIb]. Micro-propulsors have been realized using micro-technology with applications targeted at the stabilization of micro-satellites or small drones.

Today, these works are realized using energetic nano-materials. These materials have energy densities that are much greater than batteries and are capable of releasing that energy in a very short time. Therefore, they offer much higher power densities than batteries (see the Ragone diagram in Figure 4.6).

These nano-materials are comparable with hydrocarbons. Like them, they release energy in thermal form. A converter and its conversion efficiency should be considered if energy is to be recovered in an electric form.

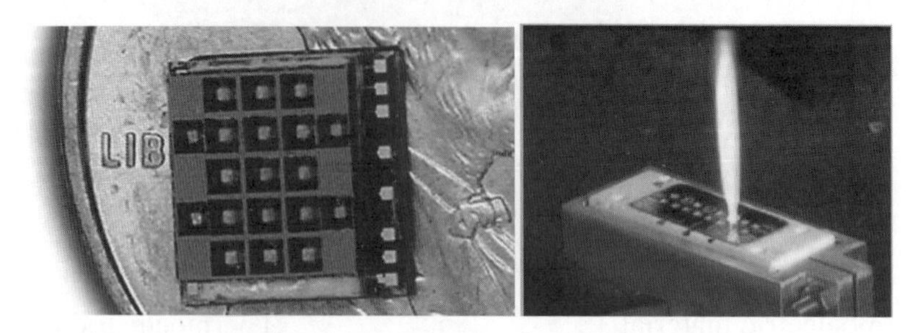

Figure 4.18. *Pyrotechnical micro-capsule die (empty), integrated lighters on the substrate [DAR]*

A review of micro-pyrotechnics is available in reference [ROS 07].

4.6. Pyroelectricity

Beyond the difference in physical principles and materials, pyroelectricity differs from thermoelectricity at the level of implementation since conversion is based on a temporal variation of temperature and not on a localized temperature gradient. Pyroelectricity is linked to piezoelectricity: charges appear after variations in temperature. Pyroelectricity is essentially used in the sensors. However, today, several research teams are exploiting pyroelectricity in order to obtain a direct conversion from thermal to electrical energy. One of the major problems for recovering energy has proved to be the slow evolution of temperatures. It would be illusory to want to recover electrical energy based on variations of daily temperatures. In that case, the leakage currents in the materials would cancel the charges produced by the pyroelectric effect. Certain integrated applications exploit an electret in order to obtain increased powers even at low frequency [SAK 08].

4.7. Tribo-electricity

A Canadian team (University of Edmonton) has published works on a very simple nano-generator that can recover electrostatic charges from water, which is in forced circulation in a network of micro-channels [YAN 03].

4.8. Radioactive source

The heat produced by radioactive materials is exploited in satellites, equipped, for example, with Stirling mini-motors. Some realizations of integrated micro-generators use radioactive sources, in order to heat a thermoelectric element [ROM 08], or to charge an electret that discharges regularly [LAL 05]. Some time ago, these sources were even used in cardiac stimulators. This technology is no longer used, but the lifespan of these sources is such that some elderly patients are still equipped with them.

4.9. Recovering ambient energy

MEMS dedicated to recovering ambient energy (*harvesting*, *scavenging*) form a very dynamic branch of the Power MEMS family, which is of interest to industry (to give an idea of the economic impact, the company Yole Development is selling its report on the state of the art of the market in March 2009 for nearly US$5,000 dollars [YOL 09]).

4.9.1. *Solar*

Photo-voltaic systems have been used since the 1970s in watches and pocket calculators, and progress in the efficiency of materials has been constant but slow. However, work on integrating solar micro-panels to supply MEMS [LEE 95; LEW 09; BER 05] is quite difficult to find as these generally use "traditional" solar technology that is developed on silicon. There are many articles on improvements of certain technological aspects of the materials or on the integration technology.

4.9.2. *Thermal*

We have previously seen that there are different ways to produce electricity from thermal energy. The conventional solutions that are used in large dimensions and based on a gradient between a hot source and a cold source can be used (*Stirling* μ-generator, etc.). It is also possible to obtain direct conversion of heat using pyroelectricity or thermoelectricity. However, in some cases, these solutions do not fulfill the required specifications. Therefore, it is possible to mix several physical principles to overcome this.

For example, the generator below is meant to produce a pulse of energy when a threshold temperature is reached. It uses magnetic and piezoelectric coupling principles: a magnet, attracted by a piece of iron-nickel, constrains a piezoelectric bimorph. The piece of iron-nickel loses its magnetic properties at a certain threshold temperature (determined by the materials composition): the magnet is no longer attracted, it suddenly relaxes the constraint in the piezoelectric. The electrical energy produced can then supply a wireless emitter that transmits information to a center of acquisition. The process is reversible during cooling [CAR 08].

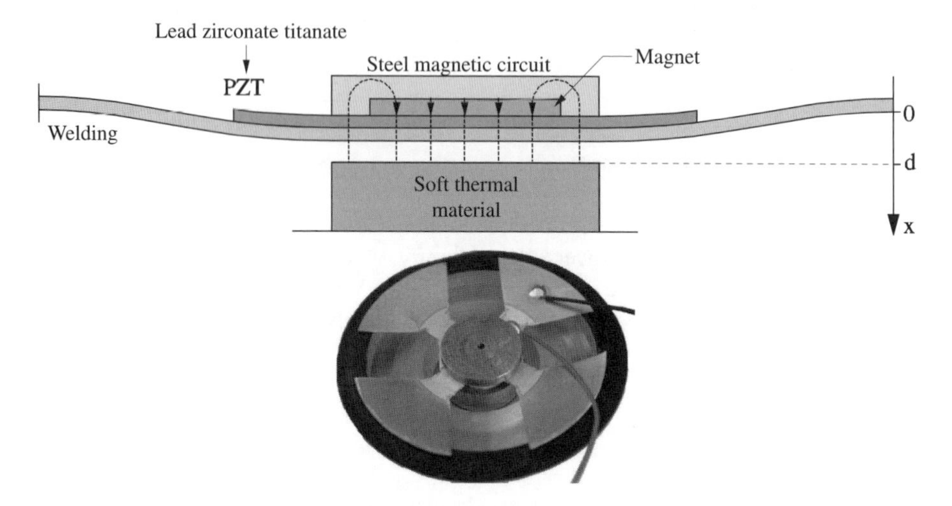

Figure 4.19. *Recovery of ambient thermal energy: magneto-piezoelectric hybrid device [CAR 08]*

4.9.3. *Chemical: living energy*

Children learn electrochemistry by planting electrodes in potatoes. Researchers, being big children, have developed scattered sensors on this principle: they are supplied by planting electrodes directly in the root of the tree and in the alkaline earth around the root [LOV 08].

Other laboratories are interested in activating a chemical cell using the addition of human urine, enabling it to be "charged" in a non-urban environment [BAN 05], or in the transformation of glucose in a fuel cell.

4.9.4. *Mechanical*

Magnetic micro-generators are naturally considered for the recovery of mechanical energy from vibrations. In the framework of the FP6 VIBES project [PROa], the Tyndall Institute (Cork, Ireland) and their colleagues in Southampton

(UK) developed a generator with a vibrating pallet on which is placed a micro-coil that can be integrated or wound from wire [KUL 08; BEE 07]. The efficiencies and the powers are weak. Other teams have also worked on electro-magnetic conversion into vibration [SAR 08; WAN 07a] or into rotation [MAR 08], but in general the magnetic solution suffers under ultra-miniaturization in the special case of recovering vibrations, because the amplitudes are small, which very much limits the variations in magnetic induction in the associated micro-coils.

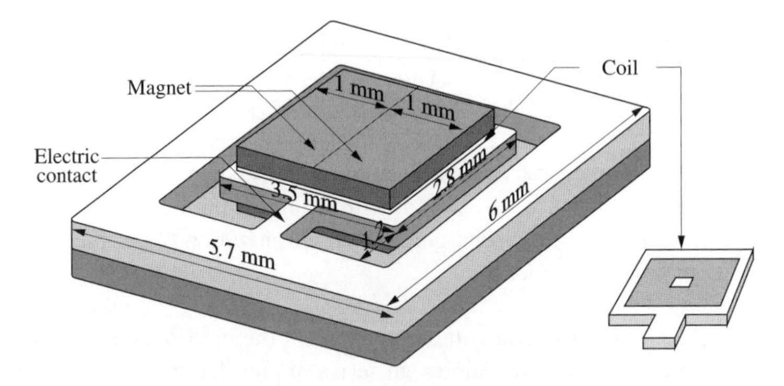

Figure 4.20. *Micro-generators recovering vibrations (VIBE) integrated on silicon: electromagnetic (integrated coils and electro-formed magnets; Tyndall Institute)*

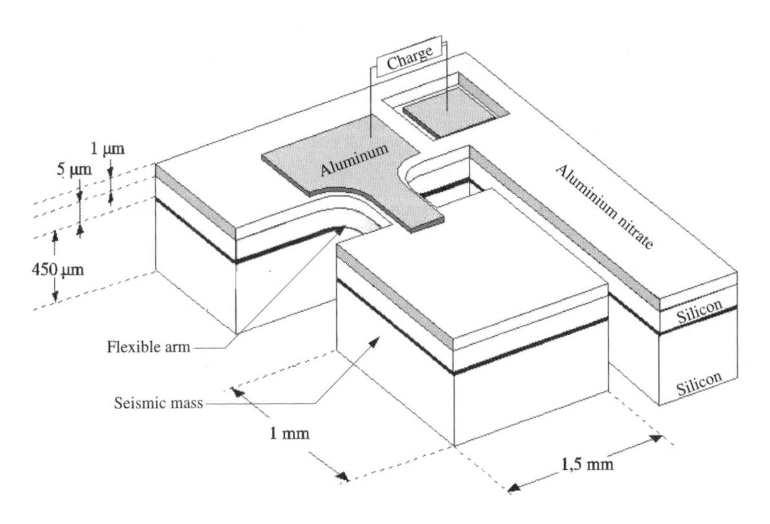

Figure 4.21. *Micro-generators recovering vibrations (VIBE) integrated on silicon: piezoelectric (TIMA)*

Within the VIBES framework, TIMA (Techniques de l'Informatique et de la Microélectronique pour l'Architecture des systems intégrés) in Grenoble has chosen a vibrating piezoelectric beam [AMM 08]. The power and voltage are weak, and the addition of electronics to amplify the voltage is necessary. Other teams are developing piezoelectric recuperators in micro- [KOK 08; RREN 08; PRI 07; XIA 06], or even nano-systems [WAN 08a] (Macroscopic piezoelectric systems are already commercialized [VOL].).

Reviews compare certain magnetic solutions [ARN 07], piezoelectric solutions [LEF 06; ANT 07; SHU 06], both at the same time [WU 08], or other solutions [MIT 07; CHA 08; PAR 05; YAN 08; MAT 05; WAN 07b].

Similar devices convert recovered vibration with the help of electrostatic devices [DES 05]. This is generally associated with charge "pumps", where a variable capacitance is charged and discharged in synchronization with the near or far position of its mobile electrodes.

In order to better exploit recovered energy from the environment, some systems link together several physical principles, in series or parallel hybrid systems (thermal + RF [WAN 08b; LHE 07], thermal + vibrations [SAT 07]). Finally, we should mention the low-voltage switch commercialized since 2005 by EnOcean [ENO] for domestic applications: a wall-mounted interrupter that transmits on/off information via radio to a receiver, which is supplied by the mechanical energy of a finger pushing on the interrupter via a macroscopic piezoelectric bimetallic strip.

4.9.5. *Transponder*

RFID chips [KAY 07] (*Radio Frequency Identification*) can be implanted in products and in buildings [SCH 03], even in patients [LU 07]. These chips function without the need for energy storage, as they use transponders that are activated over a frequency range and answer to the emitter using a signal that may be fixed (simple identification) or may depend on local conditions (remote-access sensor) [SEK 07]. RFID chips are available commercially [ATM, BLU].

4.10. Associated electronics: use of electricity – onboard EP

One factor that is related to micro-sources is the use of the electricity produced by them. Some only produce a few tens of millivolts, often fluctuating as the source dictates.

Converters should therefore be integrated within systems [XU 05; LEF 07]. As they power themselves using the energy that they must process with minimum waste, they must therefore be small, but more importantly autonomous and with low consumption [PET 08; CRE 07; RAI 07; CHA 08].

Below a certain voltage threshold, the majority of converters cannot even power themselves; the Fraunhofer Institute for Integrated Circuits (IIS) has developed a transformer for integrated voltages which is 1.5 × 1.5 mm and can function with a minimum voltage of 20 mV [PROb]. This circuit can supply voltage of a few volts with an efficiency of 30 to 80% depending on the voltage and the charge.

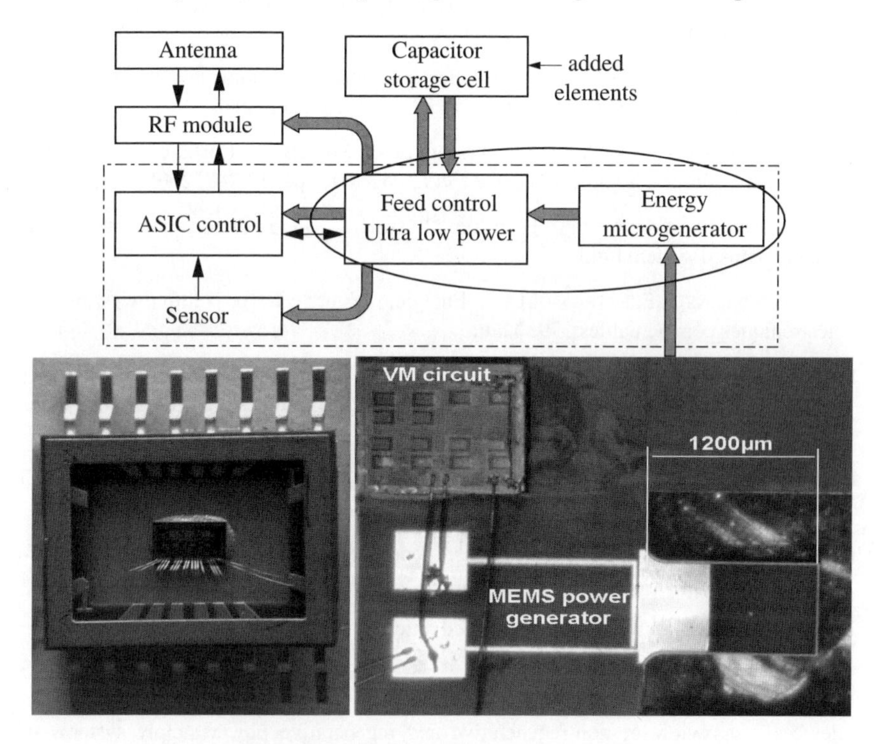

Figure 4.22. *Low consumption integrated micro-converters (G2Elab [RAI 07], TIMA [AMM 08])*

4.11. Bibliography

[AMM 08] AMMAR Y., MARZENCKI M., BASROUR S., "Integrated power harvesting system including a MEMS generator and a power management circuit", *Sens. Actuators A Phys.*, vol. 145-146, pp. 363-370, 2008.

[ANT 07] ANTON S.R. *et al.*, "A review of power harvesting using piezoelectric materials (2003–2006)", *Smart Mater. Struct.*, vol. 16, pp. R1-R21, 2007.

[ARN 07] ARNOLD D.P., "Review of microscale magnetic power generation", *IEEE Trans. Magn.*, vol. 43, no. 11, pp. 3940-3951, 2007.

[ATM] ATMEL, RF identification, June 2010, http://www.atmel.com/products/rfid/default.asp.

[BAG 08] BAGGETTO L. *et al.*, "High energy density all-solid-state batteries: a challenging concept towards 3d integration", *Adv. Funct. Mater.*, vol. 18, pp. 1057-1066, 2008.

[BAN 05] BANG LEE K., "Urine-activated paper batteries for Biosystems", *J. Micromech. Microeng.*, vol. 15, pp. S210-S214, 2005.

[BEE 07] BEEBY S.P. *et al.*, "A micro electromagnetic generator for vibration energy harvesting", *J. Micromech. Microeng.*, vol. 17, pp. 1257-1265, 2007.

[BER 05] BERMEJO S., CASTANER L., "Dynamics of MEMS electrostatic driving using a photovoltaic source", *Sens. Actuators A Phys.*, vol. 121, pp. 237-242, 2005.

[BLU] BLUECHIIP, About the Bluechiip™ system, June 2010, http://www.mems-id.com/memsid-system.html.

[BUL] BULLETINS ELECTRONIQUES, Energie, June 2010, http://www.bulletins-electroniques.com/actualites/57937.htm.

[CAR 08] CALIOZ L., DELAMARE J., BASROUR S., POULIN G., Hybridization of Magnetism and Piezoelectricity for an Energy Scavenger based on Temporal Variation of Temperature, Proceedings from DTIP'08, April 2008.

[CHA 08] CHALASANI S., CONRAD J.M., "A survey of energy harvesting sources for embedded systems", *Southeastcon*, IEEE, pp. 442-447, 3-6 April 2008.

[CHA 08] CHANDRAKASAN A.P. *et al.*, "Next generation micro-power systems, Symposium on VLSI Circuits", *Dig. Techn. Papers*, p.2-5, 2008.

[CHI 06] Chiao M. *et al.*, "Micromachined microbial and photosynthetic fuel cells", *J. Micromech. Microeng.*, vol. 16, pp. 2547-2553, 2006.

[COO 08] COOK-CHENNAULT K.A., THAMBI N., SASTRY A.M., "Powering MEMS portable devices – a review of non-regenerative and regenerative power supply systems with special emphasis on piezoelectric energy harvesting systems", *Smart Mater. Struct.*, vol. 17, no. 4, 043001, (33p.), 2008

[CRE 07] CREBIER J.C. *et al.*, "High efficiency 3-phase CMOS rectifier with step-up and regulated output voltage – design and system issues for micro-generation applications", *Proc. DTIP*, Stresa, Italy, pp. 338-343 (http://hdl.handle.net/2042/14622), 2007.

[DAN 02] DANROC J. *et al.*, "Mini et micro-batteries", *J. Phys. IV*, France 12, Pr2-121, 2002.

[DAR] DARPA, Micro-propulsion projects, June 2010, http://design.caltech.edu/micropropulsion/.

[DES 05] DESPESSE G. *et al.*, "Fabrication and characterisation of high damping electrostatic micro devices for vibration energy scavenging", *Proc. DTIP 2005 Conference (Design, Test Integration and Packaging of MEMS and MOEMS)*, pp. 386-390, 2005.

[EFT 04] EFTEKHARI A., "Fabrication of 5 V lithium rechargeable micro-battery", J. Power Sources, vol. 132, 1-2, pp. 240-243, 2004.

[ENO] ENOCEAN, Energy Harvesting, June 2010, http://www.enocean.com/en/energy-harvesting, http://radiospares-fr.rs online.com/web/0189065.html, http://www.domo-energie.com/fr/page.asp?Id=221.

[EUR 01] EUREKA, Micro fuelcells, June 2010, http://www.eureka.gme.usherb.ca/memslab/MEMSLab_f/fuelcells_f.htm.

[GEO] GEORGIA TECH – MEMS, June 2010, http://mems.mirc.gatech.edu/msma/index.htm.

[GON 06] GONDRAND C., Analyse des transferts d'eau dans les micropiles à combustible, PhD thesis, INP-Toulouse, 2006.

[HER 08a] HERRAULT F., JI C.H., ALLEN M.G., "Ultraminiaturized high-speed permanent-magnet generators for milliwatt-level power generation", *J. MEMS*, vol. 16, no. 6, pp. 1376-1387, 2008.

[HER 08b] HERRAULT F. *et al.*, "High temperature operation of multi-watt, axial-flux, permanent-magnet microgenerators", *Sens. Actuators A Phys.*, vol. 148, pp. 299-305, 2008.

[HOL 05] HOLMES A.S., HONG G., PULLEN K.R., "Axial-flux permanent magnet machines for micropower generation", *J. Microelectromech. Syst.*, vol. 14, pp. 54-62, 2005.

[JAC 02] JACQUES R., "Sources d'énergie embarquées", in CUGAT O., *Micro-actionneurs électroactifs*, pp. 243-261, Hermès, Paris, 2002.

[KAM 07] KAMEL F.E., GONON P., "Dielectric response of Cu/amorphous $BaTiO_3$/Cu capacitors", *J. Appl. Phys.*, vol. 101, 073901, 2007.

[KAM 08] KAMITANI A., MORISHITA S., KOTAKI H., ARSCOTT S., "Miniaturized micro-DMFC using silicon microsystems techniques: performances at low fuel flow rates", *J. Micromech. Microeng.*, vol. 18, 125019, 2008.

[KAN 09] KANG B., CEDER G., "Battery materials for ultrafast charging and discharging", *Nature*, vol. 458, pp. 190-193, 2009.

[KAR 08] KARPELSON M., GU-YEON W., WOOD R.J., "A review of actuation and power electronics options for flapping-wing robotic insects", *IEEE Int. Conf. Robotics and Automation ICRA*, pp. 779-786, INSPEC 10014809, May 19-23, 2008.

[KAY 07] KAYA T., KOSER H., "A new batteryless active RFID system: smart RFID",, *Proceedings from: 1st Annual RFID Eurasia Conference*, pp. 1-4, Istanbul, Turkey, September 2007.

[KIM 02] KIM J.Y., CHUNG I.J., "An all-solid-state electrochemical supercapacitor based on poly3-(4-fluorophenylthiophene) composite electrodes", *J. Electrochem. Soc.*, vol. 149, no. 10, pp. A1376-A1380, 2002.

[KOK 08] KOK S.L., WHITE N.M., HARRIS N.R., "Free-standing thick-film piezoelectric energy harvester", *Proc. IEEE Sensors 2008*, Lecce, Italy, 2008.

[KUL 08] KULKARN *et al.*, "Design, fabrication and test of integrated micro-scale vibration-based electromagnetic generator", *Sens. Actuators A Phys.*, vol. 145-146, pp. 336-342, 2008.

[KUN 07] KUNDU A. *et al.*, "Micro-fuel cells – current development and applications", *J. Power Sources*, vol. 170, no. 1, pp. 67-78, 2007.

[LAL 05] LAL A., DUGGIRALA R., LI H., "Pervasive power: a radioisotope-powered piezoelectric generator", *PERVASIVE computing*, pp. 53-60, IEEE CS and IEEE ComSoc 1536-1268/05/© 2005.

[LEE 95] LEE J.B. *et al.*, "A miniaturized high-voltage solar cell array as an electrostatic MEMS power supply", *J. MEMS*, vol. 4, no. 3, pp. 102-106, 1995.

[LEF 06] LEFEUVRE E. *et al.*, "A comparison between several vibration-powered, piezoelectric generators for standalone systems", *Sens. Actuators A Phys.*, vol. 126, pp. 405-416, 2006.

[LEF 07] LEFEUVRE E. *et al.*, "Buck-boost converter for sensorless power optimization of piezoelectric energy harvester", *IEEE Trans.*, vol. 22, no. 5, pp. 2018-2025, 2007.

[LEO 09] LEONOV V., VULLERS R.J.M., "Wearable thermoelectric generators for body-powered devices", *J. Electron. Mater.*, Special Issue Paper, available online, 2009.

[LEW 09] LEWIS J., ZHANG J., JIANG X., "Fabrication of organic solar array for applications in microelectromechanical systems", *J. Renew. Sustain. Energy*, vol. 1, no. 1, 013101, 2009.

[LHE 07] LHERMET H. *et al.*, "On chip post-processed microbattery powered with RF and thermal energy through a power management circuit", *Proc. ICICDT07, IEEE International Conference on IC Design and Technology*, Grenoble, France, June 2007.

[LIU 08] LIU R., IL CHO S., BOK LEE S., "Poly(3,4-ethylenedioxythiophene) nanotubes as electrode materials for a high-powered supercapacitor", *Nanotechnology*, vol. 19, 215710, 2008.

[LOV 08] LOVE C.J., ZHANG S., MERSHIN A., "Source of sustained voltage difference between the xylem of a potted Ficus benjamina tree and its soil", *PLoS One*; vol. 3, e2963, Aug 13 2008.

[LU 07] LU H.M. *et al.*, "MEMS-based inductively coupled RFID transponder for implantable wireless", *Sensor Applications IEEE Trans. Magn.*, vol. 43, no. 6, pp. 2412-2414, 2007.

[MAR 05] MARSACQ D., "Les micro-piles à combustible", *Clefs CEA*, no. 50/51, Winter 2004-2005.

[MAR 08] MARTINEZ-QUIJADA J., CHOWDHURY S., "A two-stator MEMS power generator for cardiac pacemakers", *IEEE Int. Symposium on Circuits and Systems*, ISCAS, Seattle, USA, pp. 161-164, 2008.

[MAT 05] MATEU L., MOLL F., "Review of energy harvesting techniques and applications for microelectronics", *Proc. SPIE Microtechnologies for the New Millenium*, Seville, Spain, pp. 359-373, 2005.

[MAT 08] MATHÚNA C.Ó., O'DONNELL T., MARTINEZ-CATALA R.V., ROHAN J. O'FLYNN B., "Energy scavenging for long-term deployable wireless sensor networks", *Mater. Today Talanta*, vol. 75, no. 3, pp. 613-623, 2008.

[MIT 07] MITCHESON P.D. *et al.*, "Performance limits of the three MEMS inertial energy generator transduction types", *J. Micromech. Microeng.*, vol. 17, pp. S211-S216, 2007.

[MOR 07] MORSE J.D., "Micro-fuel cell power sources", *Int. J. Energy Res.*, vol. 31, no. 6-7, pp. 576-602, 2007.

[NAG 04] NAGASUBRAMANIAN G., DOUGHTY D.H., "Electrical characterization of all-solid-state thin film batteries", *J. Power Sources*, vol. 136, no. 2, pp. 395-400 (http://dx.doi.org/10.1016/j.jpowsour.2004.03.019), 2004.

[NOV] NOVADEM, Micro-drones, June 2010, www.novadem.com.

[PAN 07] PAN C.T., WU T.T., "Development of a rotary electromagnetic microgenerator", *J. Micromech. Microeng.*, vol. 17, pp. 120–128, 2007.

[PAR 05] PARADISO J.A., STARNER T., "Energy scavenging for mobile and wireless electronics", *IEEE Pervasive Computing 4*, vol. 1, pp. 18-27, INSPEC: 8399352, 2005.

[PAX] PAXITECH, Portable fuel cells, June 2010, http://www.paxitech.com/.

[PEI 04] PEIRS J., REYNAERTS D., VERPLAETSEN F., "A microturbine for electric power generation", *Sens. Actuators A Phys.*, vol. 113, pp. 86–93, 2004.

[PET 08] PETERS C. *et al.*, "A CMOS integrated voltage and power efficient AC/DC converter for energy harvesting applications", *J. Micromech. Microeng*, vol. 18, 104005, 2008.

[PIC 07] PICHONAT T., GAUTHIER-MANUEL B., "Recent developments in MEMS-based miniature fuel cells", *Microsyst Technol.*, vol. 13, pp. 1671–1678, 2007.

[PRI 07] PRIYA S., "Advances in energy harvesting using low profile piezoelectric transducers", *J. Electroceramics*, vol. 19, no. 1, pp. 167-184, 2007.

[PROa] PROJET EUROPÉEN VIBES, Homepage, June 2010, http://www.vibes.ecs.soton.ac.uk/.

[PROb] PROJET FRAUNHOFER, "Nanocomposites thermoélectriques", June 2010, http://www.bulletins-electroniques.com/actualites/57936.htm.

[RAI 06] RAISIGEL H., CUGAT O., DELAMARE J., "Permanent magnet planar micro-generators", *Sens. Actuators A Phys.*, vol. 130–131, pp. 438–444, August 2006.

[RAI 07] RAISIGEL H. *et al.*, "Autonomous, low voltage, high efficiency CMOS rectifier for 3-phase micro generators", *Transducers 07/Eurosensors* XXI, pp. 883-886, Lyon, France, June 10-14, 2007.

[REN 08] RENAUD M. *et al.*, "Fabrication, modelling and characterization of MEMS piezoelectric vibration harvesters", *Sens. Actuators A Phys.*, vol. 145-146, pp. 380-386 (doi:10.1016/j.sna.2007.11.005), 2008.

[ROM 08] ROMER M. *et al.*, "Ragone plot comparison of radioisotope cells and the direct sodium borohydride/hydrogen peroxide fuel cell with chemical batteries", *IEEE Trans. Energy Conversion*, vol. 23, no. 1, pp. 171-178, 2008.

[ROS 05] ROSSI C., ESTÈVE D., "Micropyrotechnics, a new technology for making energetic microsystems: review and prospective", *Sens. Actuators A Phys.*, vol. 120, pp. 297-310 (doi:10.1016/j.sna.2005.01.025), 2005.

[ROS 07] ROSSI C. *et al.*, "Nanoenergetic materials for MEMS: a review", *J. MEMS*, vol. 16, no. 4, pp. 919-931, INSPEC 9606478, 2007.

[ROU 04] ROUNDY S. *et al.*, "Power sources for wireless sensor networks", in *Wireless Sensor Network*, pp. 1-17, Lecture Notes in Computer Science, Springer, vol. 2920, 2004.

[SAK 08] SAKANE Y., SUZUKI Y., KASAGI N., "The development of a high-performance perfluorinated polymer electret and its application to micro power generation", *J. Micromech. Microeng.*, vol. 18, 104011, 6 p., 2008.

[SAL 08] SALOT R. *et al.*, "Microbattery technology overview and associated multilayer encapsulation process", *MRS Fall Meeting*, Boston, USA, (http://www.science24.com/paper/15827), 2008.

[SAR 08] SARI I., BALKAN T., KULAH H., "An electromagnetic micro power generator for wideband environmental vibrations", *Sens. Actuators A Phys.*, vol. 145-146, pp. 405-413, 2008.

[SAT 07] SATO N. *et al.*, "Monolithic integration fabrication process of thermoelectric and vibrational devices for microelectromechanical system power generator", *Japan. J. Appl. Phys.*, vol. 46, no. 9A, pp. 6062–6067, 2007.

[SCH 03] SCHNEIDER M., Radio frequency identification (RFID) technology and its applications in the commercial construction industry, PhD thesis, Bauhaus-Universität Weimar 2003.

[SCH 08] SCHWYTER E. *et al.*, Flexible micro thermoelectric generator based on electroplated Bi2+xTe3-x, Proc. DTIP of MEMS/NEMS, DTIP'08, Nice, France (http://hal.archives-ouvertes.fr/docs/00/27/76/76/PDF/dtip08046.pdf), 2008.

[SEK 07] SEKI T. *et al.*, "SNA-MEMS batteryless-wireless sensing module utilizing RFID system", *4th Int. Conf. on RFID*, Instanbul, Turkey, pp. 243-243, 2007.

[SEN 08] SENESKY M.K., SANDERS S.R., "A millimeter-scale electric generator", *IEEE Trans. Industr. Appl.*, vol. 44, no. 4, pp. 1143-1149, 2008.

[SHE 02] SHERBROOKE UNIVERSITY, Development of a MEMS-based Rankine cycle steam turbine for power generation: project status, June 2010, http://www.eureka.gme.usherbrooke.ca/memslab/docs/PowerMEMS-Rankine-Review-paper-final.pdf.

[SHU 06] SHU Y.C. *et al.*, "Analysis of power output for piezoelectric energy harvesting systems", *Smart Mater. Struct.*, vol. 15, pp. 1499-1512, 2006.

[STR 04] STRASSER M. *et al.*, " Micromachined CMOS thermoelectric generators as on-chip power supply ", *Sens. Actuators A Phys.*, vol. 114, pp. 362–370, 2004.

[UNIa] UNIVERSITY OF BERKELEY, MEMS rotary internal combustion engine, June 2010, http://www.me.berkeley.edu/mrcl.

[UNIb] UNIVERSITY OF BERKELEY, MEMS Rockets, June 2010, http://www.me.berkeley.edu/mrcl/rockets.html.

[UNIc] UNIVERSITY OF MINNESOTA, Micro-homogenous charge compression ignition (HCCI) combustion: Investigations employing detailed chemical kinetic modeling and experiments, June 2010, http://www.menet.umn.edu/~haich/paper2.pdf.

[UNId] UNIVERSITY OF BIRMINGHAM, Mirco-engineering and Nano-technology Research Group, June 2010, http://www.micro-nano.bham.ac.uk/micro.htm.

[VOL] VOLTURE, MIDE, Vibration energy harvesting products, June 2010, http://www.mide.com/products/volture/volture_catalog.php.

[WAN 05] WANG W. *et al.*, "A new type of low power thermoelectric micro-generator fabricated by nanowire array thermoelectric material", *Microelectron. Eng.*, vol. 77, pp. 223–229, 2005.

[WAN 07a] WANG P.H. *et al.*, "Design, fabrication and performance of a new vibration-based electromagnetic micro power generator", *Microelectron. J.*, vol. 38, pp. 1175–1180, 2007.

[WAN 07b] WANG L., YUAN F.G., "Energy harvesting by magnetostrictive material (MsM) for powering wireless sensors in SHM", in *SPIE Smart Structures and Materials & NDE and Health Monitoring, 14th International Symposium* (SSN07), March 18-22, 2007.

[WAN 08] WANG Z.L., "Energy harvesting for self-powered nanosystems", *Nano. Res.*, vol. 1, p. 1-8, 2008 (see also http://www.technovelgy.com/ct/Science-Fiction-News.asp?NewsNum=1000).

[WU 08] WU X., KHALIGH A., XU Y., "Modeling, design and optimization of hybrid electromagnetic and piezoelectric MEMS energy scavengers", *Proc. IEEE 2008 Custom Intergrated Circuits Conference (CICC)*, San José, California, USA, pp. 177-180, 2008.

[XIA 06] XIA Y.X., Self-powered wireless sensor system using MEMS piezoelectric micro power generator (PMPG), PhD thesis, MIT, 2006.

[XU 05] XU S.W. *et al.*, "Converter and controller for micro-power energy harvesting", *Proc. IEEE Applied Power Electronics* APEC 2005, Austin, Texas, USA, vol. 1, pp. 226-230, 2005.

[YAN 03] YANG J. *et al.*, "Electrokinetic microchannel battery by means of electrokinetic and microfluidic phenomena", *J. Micromech. Microeng.*, vol. 13, pp. 963-970, 2003.

[YAN 07] YANG Y., WEY X.J., LIU J., "Suitability of a thermoelectric power generator for implantable medical electronic devices", *J. Phys. D: Appl. Phys.*, vol. 40, pp. 5790–5800, 2007.

[YAN 08] YANQIU L. *et al.*, "Hybrid micropower source for wireless sensor network", *IEEE Sensors J.*, vol. 8, no. 6, pp. 678-681, 2008.

[YEA 07] YEATMAN E.M., "Applications of MEMS in power sources and circuits", *J. Micromech. Microeng.*, vol. 17, pp. S184-S188, 2007.

[YOL 09] YOLE DEVELOPMENT, "MEMS energy harvesting devices", *Technologies and Markets*, http://www.yole.fr/pagesAn/products/MEMS_Energy_Harvesting.asp, March 2009, accessed June 2010.

[ZWY 06] ZWYSSIG C., KOLAR J.W., "Design considerations and experimental results of a 100 W, 500,000 rpm electrical generator", *J. Micromech. Microeng.*, vol. 16, pp. S297–S302, 2006.

Chapter 5

Hydrogen Storage

5.1. Introduction

Hydrogen is the most abundant element, and along with carbon and oxygen it forms the basis of chemical reactions that are easy to set up and that can provide energy. Hydrogen is recognized as the energy vehicle of the future taking the place of carbon and hydrocarbons, for which natural resources are almost exhausted or else difficult to access.

Moreover, the use of hydrogen as a clean energy vehicle is considered the only reasonable long-term possibility, as its combustion only produces water, and this should help to combat the greenhouse effect, which is linked to the level of carbon dioxide (CO_2) in the atmosphere, which is currently increasing measurably due to human activities. Therefore, hydrogen is being called upon to play a major role in energy for centuries to come by taking the place of fossil fuels, which are currently dominant due to their abundance, their ease of storage and use, and their initial cost. It is these criteria on which hydrogen needs to become competitive during a period of transition, research, and optimization. However, its development will occur in parallel with competitors, as the needs and applications of energy are so varied.

The dawn of the first industrial and economic era of hydrogen is still faced with significant technological obstacles across the four energy transfer steps: production, storage, distribution, and use (PSDU). Due to their evident interdependence, answers should be determined for their global merits in the PSDU system. Hydrogen from different possible modes of production will not be stored in the same way, and

---

Chapter written by Daniel FRUCHART.

different chemical or physical forms of stored hydrogen will evidently not be distributable in the same way. Similarly, slow or rapid combustion of hydrogen requires very different technologies.

Regarding storage, important progress has been made over the last few years regarding the different physical or chemical states of hydrogen, in gaseous, liquid or solid form (depending on whether bonds are atomic or molecular). Many parameters come into play, but it is easy to list several important criteria such as density, specific density, sizing of the application, rapidity of reaction, reversibility, security and the global economic impact. There does not seem to be a unique solution, but a large range of possibilities and of systems.

Before proceeding to a general review of the state of the art, we will recall the essential physical, thermodynamic, and chemical characteristics for different forms of hydrogen, which will affect all the steps for putting hydrogen to work. In the following three sections we will present the most recent propositions regarding storage:

– in gaseous form;

– in liquid form;

– in solid form (bonded form).

The advantages and disadvantages of each of these methods will be discussed, with reference to the PSDU steps.

5.2. Generalities regarding hydrogen storage

5.2.1. *Pertinent energy parameters*

Fuel	Specific energy density (MJ/kg)	Energy density (MJ/l)
Hydrogen	142	8 (under 70 MPa)
Natural gas	54	10 (under 20 MPa)
Petrol	42	28

Table 5.1. *Pertinent energy parameters for* H_2

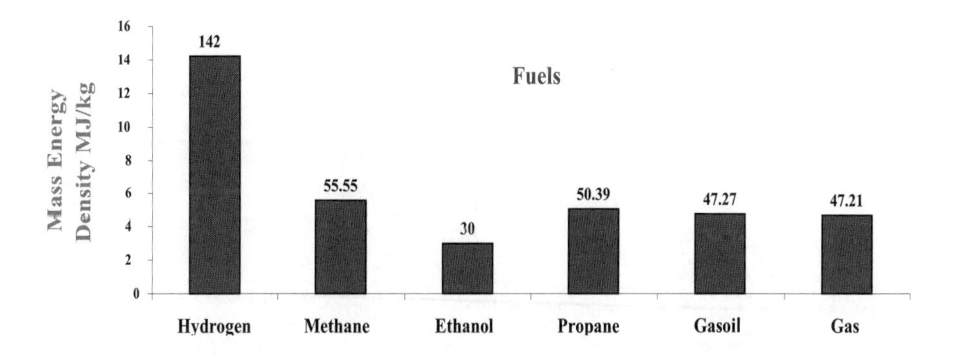

Figure 5.1. *Specific densities of different fuels*

Table 5.1 and Figure 5.1 compare the energy performance of hydrogen with that of traditional fuels. The marked advantage of hydrogen comes from the fact that the oxidation of carbon (combustion) produces four-times less energy than that of hydrogen, and that fossil fuels also produce CO_2 when oxidized. It is also important to consider the application, for example, whether hydrogen combustion leads to production of water in vapor or liquid form.

The data given above refer to the most favorable case (liquid water), and in the other case the specific energy density of hydrogen is reduced to ~125 MJ/kg H_2, which means 1 kg H_2 = 33.33 kWh.

However, the given data do not show the main disadvantage for hydrogen, compared with traditional fuels, which is its very low density: 1 kg of hydrogen occupies a volume of more than 11 m^3. As a result, it is important to be able to compress hydrogen during storage.

5.2.2. *Density versus specific density*

Figure 5.2 illustrates the volume and mass for three main forms of compressed hydrogen: physical and molecular form (i.e. high compression of the gas or cryo-liquefication), solid or "atomic" form with formation of the metal hydrides (chemical form), or by absorption into other materials (molecular form).

In the case of physical storage (under pressure or cryogenic), the mass of the barrel or the cryogenic vessel are not considered in Figures 5.2 and 5.3. These parameters are important in applications, and will be discussed in the sections relevant to each storage mode.

Volume of hydrogen stored
Loaded tank: 5 kg H_2 = 300 km

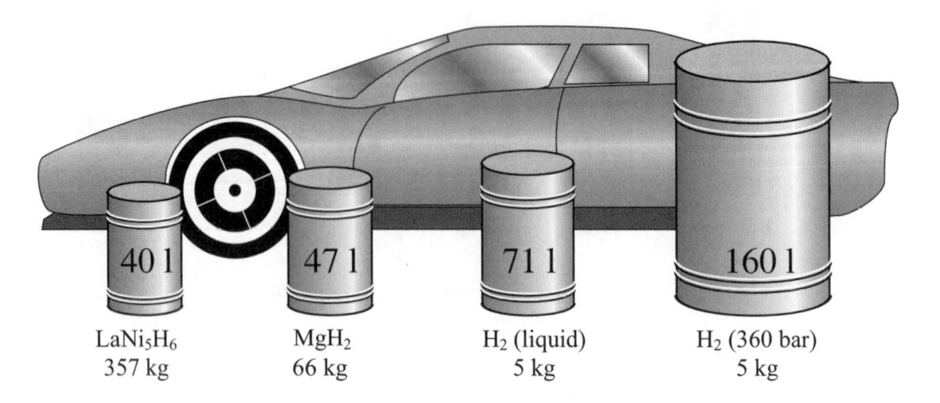

Figure 5.2. *Stored densities (source: A. Zuettel and L. Schlapbach, CNRS Grenoble)*

Type of storage		Volume	Mass (*)	Pressure	Temperature	System
	Molecular H_2	max. 33 kg $H_2 \cdot m^{-3}$	13 mass %	800 bar	298 K	Composite cylind. established
	Molecular H_2	71 kg $H_2 \cdot m^{-3}$	100 mass %	1 bar	21 K	Liquid hydrogen
	Molecular H_2	20 kg $H_2 \cdot m^{-3}$	4 mass %	70 bar	65 K	Physisorption
	Atomic H	max. 150 kg $H_2 \cdot m^{-3}$	2 mass %	1 bar	298 K	Metal hybrides
	Atomic H	150 kg $H_2 \cdot m^{-3}$	18 mass %	1 bar	298 K	Complex hybrides reversibility ?
	Atomic H	> 100 kg $H_2 \cdot m^{-3}$	14 mass %	1 bar	298 K	Alkali + H_2O

Figure 5.3. *Comparison of different types of H_2 storage*

Figure 5.3 describes the advantages and disadvantages of the different types of hydrogen storage. For a fair number of metal hydrides where absorption and desorption are effective at low pressure, the chemical reaction of desorption is an endothermic reaction, which can be considered a security feature (in case of a leak, for example).

In Figure 5.4, only the material capacities are shown, without accounting for the mass of the reservoir, or the mass of water necessary to produce hydrogen in the case of some composites such as the boronates. Note that in that figure, the intermediary objectives between 2010 and 2015 include only one "traditional" metal hydride, and beyond 2015 only boronates, alanates, and other complex hydrides are favored by the experts of the US Department of Energy (DoE). The chemical or physical risks are not included and no reference is made to the cost (of extraction or production) or even to the natural occurrence of the elements listed. Therefore, it is difficult to use such a table directly and to make it a realistic basis for applications and short-term objectives. Therefore, its use should be limited to comparisons of capacity.

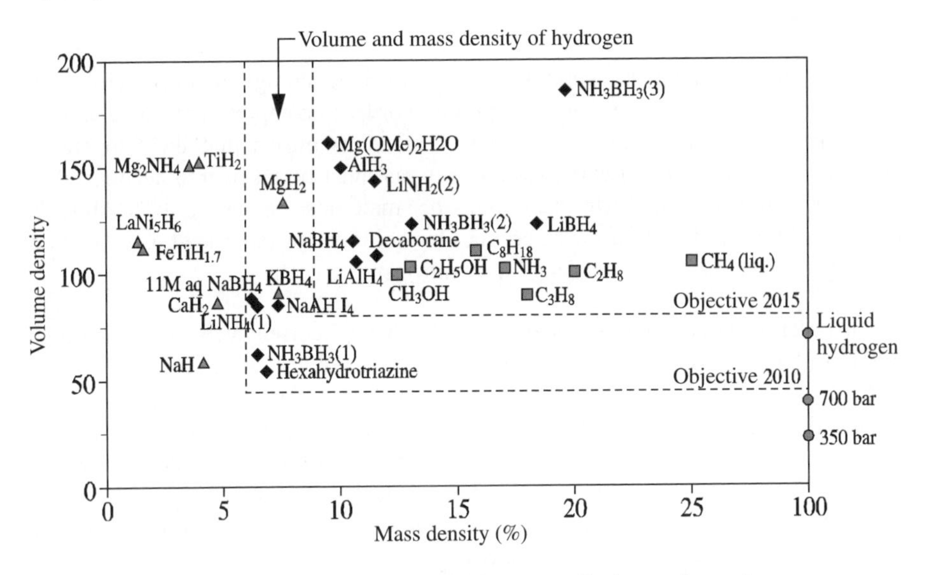

Figure 5.4. *Densities and specific densities of hydrogen for various energy sources (Source DoE)*

5.3. Pressurized storage

5.3.1. *Reservoirs*

Pressurized storage is a technique that has been tested in industry for many years, for under 200 bars with distribution using steel tubes (~ 10 Nm^3). However, distribution of very large quantities of hydrogen in traditional tubes by ground transport is not economically foreseeable. Therefore, the development of special reservoirs for mobile applications, for working under 350 bars and then under 700 bars has been undertaken by numerous national institutes, and also by the world leaders in production and distribution of gas.

The design of these reservoirs has been uniform, comprising:

– an exterior envelope with high mechanical resistance to corrosive environments (acids, for example), which is also very light; it is constructed from a composite reinforced with high-quality carbon fiber;

– an interior envelope, or "liner", made of either polymer or light metal (aluminum, for example), which are among the most efficient materials in terms of water/air-tightness. Many tests of resistance and security have been undertaken, especially with regard to mechanical behavior and behavior at explosion, in order to specify norms.

There is much interest, energywise, in pressurized storage, as in principle only 9% to 13% of the primary energy is required in order to compress the gas to 350 and 700 bars, respectively. However, filling a reservoir means that the temperature elevation inherent in rapid compression of a gas must be taken into account. As a result, the mechanical performance of the materials beyond 200°C must be considered during design of the reservoir. One solution for rapid loading consists of cooling the gaseous hydrogen using liquid nitrogen.

In order to have a range that is close to that of current vehicles, i.e. around 500 km, the reservoir will have to have a minimal volume of around 125 liters in order to supply a fuel cell (FC) and of around 250 liters to supply an internal combustion engine (ICE), with volumes not taking into account the envelopes nor the auxiliary systems. Economically, it is estimated that for the next decade the cost of a mobile pressurized reservoir should become established at less than €60 per unit assuming mass production. The cost of recycling the composite materials that make up the envelope still has to be established.

In the case of stationary applications, installations with large capacity supporting high-pressure gas (50 to 180 bars) have already been developed, essentially for the needs of the gas production and distribution industries. The cylindrical form is usually adopted (for example, 4,500 m^3, with a length of around 20 m, and 50 bars of pressure) or alternatively, the spherical form is used. In order to reduce risks, the EU standards focus on an installation pressure limited to 50 bars and a maximum volume of 350 m^3, equivalent to 400 kg of stored hydrogen.

5.3.2. *Networks*

Industrial hydrogen is distributed at lower pressure in pipelines, and the most important such network in the world is installed between France, Germany, Belgium, and Holland. Such networks could be considered to constitute "reservoirs" of a reasonable size to serve as buffer volumes between the hydrogen production ensured by intermittent sources (wind power, photovoltaics, etc.) and consumption over a relatively short period (24 to 72 hours).

Regarding underground gas storage, large-scale practical experiments (up to more than 300 Mm^3) have been carried out in France, Germany, the United Kingdom, and the United States. Finding impenetrable underground cavities is an important task and in some cases, it has not been possible to recuperate more than 50% of the gas that was injected.

5.4. Cryogenic storage

The liquid H_2 energy vehicle offers certain advantages in manipulation from production to distribution, for example, by tanker. As shown in Figures 5.3 and 5.4, the specific density is greater for the liquid than the gas, even when compressed to 700 bars. However, the first handicap in terms of efficiency for this form of hydrogen comes with the energy necessary for liquefaction, which can vary from 30 to 40% of the primary energy available depending on the technique. Secondly, losses by evaporation during storage should be taken into account, which are estimated to measure between 0.1 to 4% per day, depending on the sizing and the application. Regarding security, certain risks exist when considering the performance of containers (special steels that do not become brittle at low temperatures or with permeation of hydrogen in the walls) and the manipulation of a cryogenic fluid.

5.4.1. *Mobile storage of liquid hydrogen*

Packaging of the liquid for mobile applications, with metallic reservoirs or cryostats with two walls, separated by a vacuum or (and) with thermally isolating materials (super-isolating, perlite, etc.), was proposed by large European industrial companies (unlike in the USA or Japan). Cryogenic reservoirs have been trialed *onboard*, carrying 5 kg (for FCs) to 8 kg (for ICE) of liquid hydrogen, and ensuring, depending on the type of vehicle, a range in the order of 300 km.

Cryogenic systems constructed initially with completely metallic (internal and external) walls could weigh up to 150 kg; the substantial reduction in mass obtained by using composite materials will enable a better integration of this solution (much less volume) for propulsion of vehicles. It is difficult to forecast the price of cryogenic reservoirs – they may become less expensive if the least expensive thermal isolators are used and if regeneration of the evaporate is not considered.

Stations for distribution of liquid hydrogen have been installed, experimentally. This type of storage requires auxiliary equipment and costly security, such as connectors, an anti-high-pressure vent, and eventually repacking of the evaporated gas.

5.4.2. *Static storage of liquid hydrogen*

This type of storage is usual for all distributors of liquefied gases. As an example of liquid hydrogen storage on a very large scale, we cite the reservoir installed by NASA at the Kennedy Space Center in Florida as the only significant current use of hydrogen for a mobile application. The spherical reservoir of 20 meters diameter offers a volume of 3,800 m^3 and a specific capacity of nearly 250 metric tons of liquid hydrogen. The daily level of losses is measured to be between 0.1 and 1% of the contents. There would be no technological difficulty in developing such reservoirs to contain up to 1,000 metric tons of liquid. However, for security reasons, especially in Europe, such installations are not planned, even on smaller scales, as an explosion involving liquid hydrogen could have devastating and extensive consequences.

5.5. Solid storage

5.5.1. *Physical storage by physi-sorption (or chemi-sorption)*

Porous materials such as "activated" charcoal are known for their excellent capacity to absorb molecular or atomic gases by the principle of Van der Waals (VdeW) forces. We first of all present the carbonated materials and then other families of porous materials that can be used in these applications.

5.5.1.1. *Porous materials based on carbon*

Other than activated charcoal, numerous forms of porous or nanostructured carbons have been analyzed and are given as potential high performers or even extremely high performers (fullerenes, nanotubes – single or multi wall nanotube: SWNT, MWNT, "fish-bone carbon", cones, etc. There is even talk of a 75% concentration for use in storage!). It has been necessary to distinguish between VdeW forces between molecules and the chemical bonds that are possible for (often metallic) residual aggregates, which serve to catalyze the growth of carbon nanostructures. Ever since the 2000s, the performances announced have taken on modest values (particularly those announced by specialists in metal hydrides), especially in the ambient temperature environment.

Now it can be said that the optimized carbon nanostructures of the SWNT type perform best at ambient conditions and under normal pressure, over a range of 1 to 2% concentration. Moreover, the physical bond does not lead to thermal desorption, which can occur over a large temperature range and even for temperatures that are much higher than ambient temperature. The prohibitive cost of refined carbon nanostructures remains a major obstacle to the practical development of their application in storage.

Other extremely porous (more than 2,000 m^2/g) nanographite structures, with or without catalyzing particles inserted, are now being considered. The porosity of these structures could even be improved by mechanical procedures such as *ball milling* (BM), or mechano-synthesis. The reported performance at ambient temperature tends to progress to between 2 to 3% concentration, and the physisorption at low temperatures (77 K for example) leads to already very attractive results, at more than 7% concentration under 50 bars and around 10% concentration under 100 bars. This solution is interesting because of the compression of the H_2 gas. However, it should be remembered that these nanographites should be kept at the temperature of liquid nitrogen (in order to avoid losing their charge to molecules of H_2), and that it is necessary to charge them under pressure, under 100 bars. The ideal energy score is therefore reduced by 10% at best, two-thirds of this reduction being due to the cooling to 77 K and one-third being due to the compression of hydrogen. It is also necessary to heat the cryogenic reservoir, according to need, and of course to add systems for controlling temperature and pressure.

For large-scale applications, storage at low temperature using nanocarbons remains less energy efficient than liquid hydrogen, but has the advantage of less (around three-times less) primary energy consumption during storage. But security is also a problematic factor, as the pressure and low temperatures applied to the reservoir envelope, which will be difficult to construct, must be reconciled. The industrial cost of activated charcoal should also be considered, having a current value of around €70/kg.

5.5.1.2. *Molecular materials and other physisorbants*

Compared to the preceding materials, nanotubes and other nanostructures based on boron nitrides have not given better results, but are less economical.

Aerogels, such as those of silica, are of interest due to their low price. However, progress remains to be made as to performance based on limited specific surfaces

(1,000 m^2/g). Certain zeolites, which are thermally robust materials of low price, are also of interest, but their performance remains modest, achieving 2 to 2.5% concentration at most. Vitreous microspheres comprise a third category of material that is cheap *a priori*. The microspheres are saturated under high-pressure hydrogen and then brought back to ambient conditions, with hydrogen being recuperated thermally. A major question is how to control the activation of this type of microreservoir.

We can also list diverse materials, such as porous metallic amorphous materials and hydride slurries (mixtures of fine particles of metal alkalis with complex oils); these solutions are of average efficiency but could be useful for small-scale static storage installations, depending on their durability and their economic impact, which have yet to be determined.

A recent class of materials of potential interest are *molecular organic frameworks* (MOFs), which are large structural entities formed from ligands and organic complex radicals, articulated on metal ions of different types. These materials are crystallized with very large meshes that form large structural cages. Enormous specific surfaces (>8,000 m^2/g) are offered by these cavities, which enable physisorption of hydrogen molecules, and MOF can also form chemical bonds. Some of the constituent molecules, such as metallic oxides, are cheap. The reported performances are high (>8% concentration) but as the process is essentially physisorption, the methods are similar to those for nanocarbons, i.e. trapping hydrogen at low temperature. There is active research currently being undertaken on numerous materials of this type, but long-term performance remains to be demonstrated, in particular, when it comes to charge/discharge, whether there are any risks, and the biocompatibility of such molecules.

5.5.2. *Chemical storage*

By chemical storage, we mean the formation of compounds or hydrides where the hydrogen atom, issued by dissociation of the H_2 molecule, forms a metallic or iono-covalent bond with elements from an existing structure, most often mainly composed of metal atoms. We can distinguish metal hydrides – heavy or light, at high or low temperature – complex hydrides including alanates (formed from alkaline ions and aluminum), poly-element systems based on transition metals and alkaline earth metals, and new materials known as imides or amides where nitrogen bonds with hydrogen.

For metallic materials, absorption and desorption are essentially one-step chemical reactions (see Figure 5.5), which leave the metallic grid generally unperturbed; for complex materials, the insertion/removal of hydrogen atoms proceeds in several steps where the chemical reactions have their own energy. A common point for all the different hydrides is that most of the time the dispersion of additive particles known as “catalysts” is an operation that is necessary in order to obtain compatible kinetics for applications. With this aim, such crystallized materials are ground into powder, in order to increase their specific surfaces, using techniques such as BM.

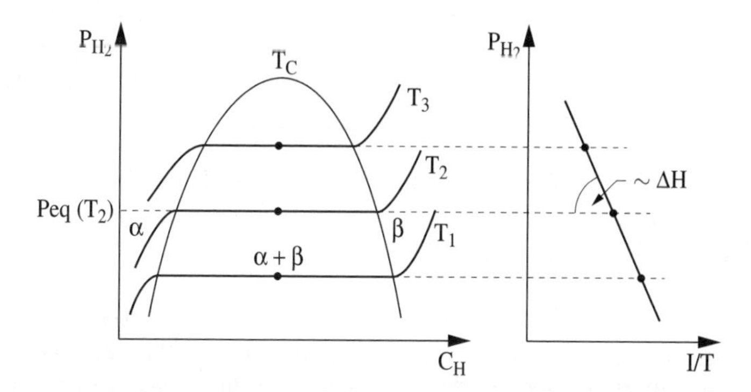

Figure 5.5. *Left: phase diagram for metal/hydrogen, relating the concentration C_H to the pressure of H_2 for different temperatures. Right: analysis of the equilibrium plateau "solid solution (α) – hydride (β)" in 1/T determining the enthalpy of formation of the hydride (Van't Hoff's law)*

A metal hydride is a chemical compound that is formed according to the following reaction:

$$M + x H2 = MHx/2 \quad [5.1]$$

by a bond between the two elements, M and H. In the hydride the H atom occupies interstitial sites defined by a certain number of neighboring M atoms. The reaction depends on the temperature and the pressure, and can be quantified by defining a group of isotherm curves, from which it is possible to extract the thermodynamic values corresponding to the heat of formation such as enthalpy, ΔH (gradient of the isotherms for 1/T), and entropy ΔS, of reaction. During desorption, hydrogen atoms leave the interstitial sites to reform molecules of H_2 gas.

5.5.2.1. *Metal hydrides*

Metal hydrides, which we can call “traditional hydrides” with reference to the above, can be classed as several families according to their structure and thermodynamic properties. We will consider the principal types:

– $LaNi_5$ (hexagonal) this forms the essential constituent of rechargeable battery electrodes, commonly known as NiMH (nickel-metal hydride);

– FeTi and derivatives (structure type CsCl);

– compounds with Laves phase structures, with formula AB_2, where A = Zr, Ti and B = a transition metal, which may be cubic or hexagonal, which in some cases are also used as battery electrodes;

– BCC structures, i.e. alloys with a centered cubic structure such as β-Ti, V, Cr, etc. which are based on these elements and MgH_2 in the form of powdered nanostructures, and doped with a hydrogenation catalyst and intermetallic compounds that are rich in magnesium.

These four types of intermetallic compounds form low-temperature hydrides that are reversible at temperatures near ambient temperature, at rather moderate pressures, generally between 1 and 10 bars.

$LaNi_5$ (with theoretical specific capacity of 1.4%) and its derivatives, formed by substitution of lanthane, is therefore very adaptable for different applications. It has been tested for different uses: on the scale of several hundreds of kilograms for the PEMFC collective (collective Proton Exchange Membrane Fuel Cell, Electricité de France (EDF)), for the propulsion of a hybrid vehicle in the United States (requiring nearly 400 kg for a range of less than 300 km), or as a mini-reservoir for integrated PEMFCs. In practice, reversibility occurs around 1% concentration, and the second drawback of this excellent hydride is the cost of the metal constituents and their weight in mobile applications.

Compounds from the FeTi family have not shown the best practical performance (≤1% concentration); however, the low price of alloys based on ferro-titaniums produced in the steel industry opens the door to large-scale (static) applications or even mobile applications with the example of submarines (Germany) bringing around 160 metric tons of such alloys to supply hydrogen to two PEMFCs of 120 kW each.

Alloys with Laves phase structures (type ZrM_2 where M = Mn, Fe, Co, Ni, etc., in various proportions) are interesting as they adapt well to ambient temperature and pressure. Their theoretical specific capacity does not exceed 2% (less than 1.5% concentration in practice) with a propensity to become unmixed during prolonged cycles. The cost of different constituents also places a limit on their use in static applications.

Activated magnesium, and to a lesser extent Mg_2Ni type compounds, have the maximum specific capacities at 7.6 and 3.9%, respectively. Their very mediocre kinetics of absorption/desorption have been the subject of intensive studies on nanostructurization by BM. With the addition of "catalyzing" additives the practical specific capacity is reduced to 6 and 6.5% at best. Pilot industrial production confirms these values. An inherent thermodynamic disadvantage is that the dissociation temperatures of these hydrides are around 250°C and 320°C, respectively, for an equilibrium pressure of 1 bar. For MgH_2 this represents around a quarter of the primary energy of hydrogen for the thermal and thermodynamic formation of the hydride. As Mg_2NiH_4 is unstable in the absorption/desorption cycle, only MgH_2 seems to be a robust hydride, it is cheap, recyclable, biocompatible, and can be mass-produced industrially. Progress in activation by metallurgic methods suggests that it will soon be possible to go beyond 7% and even to approach the theoretical reversible specific capacity (7.6% concentration). The application of such a hydride will have to be accompanied by systems that control and regenerate heat depending on the application (SOFC or ICE), in order to manage and globally optimize the energy of the system. Note that the thermal transfer parameters for a hydride have been considerably improved thanks to MgH_2-graphite composites.

5.5.2.2. *Complex hydrides*

Complex hydrides based on magnesium, calcium, or other alkaline earth elements, or rare earth elements, are being studied currently. In addition to their complexity at synthesis, they have shown no real weight performance to encourage their application, except as microreservoirs functioning in normal conditions in an integrated application.

The class of materials known as alanates, combine a very electropositive alkaline element such as A = Li, Na, K with aluminum ($AAlH_4$). This allows hope for high specific storage levels (10.5, 7.5, and 5.7%, respectively). Intensive fundamental and practical research (studies of reservoirs of several kilograms) is being undertaken in this area, taking into account the potential applications of such materials. $KAlH_4$ reacts at 300°C for absorption as well as desorption, but the situation seems more favorable with $NaAlH_4$ at 160°C and 130°C, respectively. In fact, due to the double reaction (equation II), which totally liberated the stored hydrogen, only $NaAlH_4$ appears practically cyclable and reversible, at least for a limited number of cycles:

$$NaAlH_4 \leftrightarrow 1/3\ Na_3AlH_6 + 2/3\ Al + H_2\uparrow \leftrightarrow NaH + Al + H_2\uparrow \qquad [5.2]$$

(5.5% concentration giving 3-4% hydrogen!)

Good activation kinetics are obtained with the addition of “catalysts” (such as titanium salts) via BM and nanostructurization, but alanates have been shown to be very unstable and pyrophoric and the current tendency is to be rather prudent in putting them to practical use, as absorption of hydrogen requires pressures of hydrogen of several tens of bars, even more than 100 bars.

Ternary compounds known as amides or imides have multi-step reactions that are chemically similar to total desorption (absorption). Materials such as $LiNH_2$,

$Mg(NH_2)_2$, Li-Mg-N-H, $Mg(NH_3)_6Cl_2$, and $(CH_3)_4NBH_4$ (at 18% of the theoretical mass) are effectively in characterization phase. An immediate disadvantage is that the reversibility of the cycle becomes impossible if even a trace of ammonia (NH_3) is formed during different stages of decomposition.

Other systems are also being studied, and we cannot cite them all here. Prohibitive factors may be the pressure of hydrogen necessary at recharge, or simply the temperature of the decomposition reaction that liberates all the hydrogen which remains very high (400 to 600°C). It also remains to be shown that certain compounds are chemically harmless.

5.6. Other modes of storage

This section presents less direct solutions or systems, such as non-reversible systems, chemical mixture systems, or chemical-physical hybrid modes of storage.

5.6.1. *Boronates*

Boronates or borohydrides ($LiBH_4$, $NaBH_4$, etc., $Ca(BH_4)_2$, etc.), whose formulae resemble those of alanates, have been classed apart because these materials have two forms of destabilization. One form is thermal, but then the decomposition temperature proves to be too high to be practical (min. >400°C). The other form is by hydrolysis, a solution that was already recognized in experiments by car manufacturers to supply a fuel cell in the early 2000s. In addition, integrated reservoirs to supply mobile systems have been proposed in the form of "disposable" cartridges.

According to the second mode of hydrogen formation, the sodium compound $NaBH_4$ (theoretically at 15.6% concentration) is the only driver in a kinetic reaction such as:

$$NaBH_4 + 2H_2O \leftrightarrow 4\ H_2\uparrow + NaBO_4\ (\text{aqueous}) \qquad [5.3]$$

As the reaction is exothermic, in principle it should not need any extra energy. It is possible to realize (partially) hydrolysis, but due to the kinetics, it is necessary for the reaction to take place between 140°C and 180°C. This reaction seemed easy to control using a saturated solution of borax, which is stable, and the addition of a catalyst provokes the decomposition of the boronate, in principle, according to the above equation. In fact, the reaction remains incomplete as it does not end in the formation of the oxide but rather the stabilization of the hydroxide, and therefore, the practical yield is reduced to about half. The reaction occurs at above 5% concentration. It is important that this chemical regeneration, when carried out in mass production, should not lead to the formation of boron, which is very toxic.

5.6.2. *Boronate/hydride mixtures*

Another approach proposes to combine boronates with low-temperature hydrides such as MgH_2, CeH_2, etc., in order to bring down the global decomposition temperature of the mixture, and this was recently studied with the hope of being able to use these materials with a high concentration of hydrogen. Reactions have been analyzed around 400°C, but the reactions are multi-step and complex and are difficult to control and incomplete, and for the moment they lead to fairly mediocre final results.

5.6.3. *Hybrid storage*

Confining the best "heavy" hydrides, i.e. BCC alloys, using pressurized reservoirs at intermediate pressures (for example, 200 to 350 bars) allows them to be saturated to their limit capacity, which is close to 3.7%. The "dead" volume necessarily left between the grains is used to store H_2 gas at high pressure. It is not possible to over-compress an intermetallic compound in a reservoir, as the increase in volume at hydrogenation can induce tensions that could harm the envelope. In this way, hybrid storage can be realized which, at 6% concentration, would be viable for some car manufacturers. The permanence of the alloy and the mechanical performance of the container and contents together, when subjected to mechanical and thermal strains in repeated cycles, remain to be tested.

5.7. Discussion: technical/energy/economic aspects

The review of methods for hydrogen storage leads to a few conclusions regarding maturity and applicability.

On the technological front, there is no unique hydrogen storage mode that could satisfy the diverse energy demands that are met economically and adaptably by liquid or gaseous hydrocarbons. This applies for all applications: mobile, static, or integrated.

Storage of energy resulting from new but intermittent ecological sources (wind, solar, etc.) remains one the major problems, and for a large part of current and future needs the hydrogen solution cannot be ignored either practically or economically. Table 5.2 compares the modes of hydrogen storage with modes of electrochemical storage already applied or envisaged for cars.

It turns out that only liquid hydrogen technology satisfies the criteria (6-9% concentration/kg of system, and 45-80 kg H_2/m^3 of a full tank system) given by the

DoE for the 2010-2015 period and for mobile applications. This track is also industrially exploited for large volume units (stationary storage, transport by containers). However, the energy balance sheet for storage is quite poor, with regard to liquefaction, and moreover the impact of an accident is far from negligible.

Storage mode	**Wh/kg**	**Wh/l**
Batteries		
Lead	30	70
NiMH	70	175
Li-metal-ion	100	200
Compressed H_2		
350 bar	2,000	700
700 bar	1,666	1,165
Liquid H_2	1,885	1,400
Hydrides		
Low temperature	535	2,000
High temperature	1,880	1,600
Activated charcoal	2,000	1,000
Hydrocarbon	11,660	8,750

Table 5.2. *Comparison of energy densities of batteries and hydrogen*

For some, "high temperature" hydrides present the properties required by the DoE; however, technological efforts still have to be made before mass production can take place. The energy balance sheet, with regard to enthalpy of formation, is a bit better than for the previous solution. The security factor seems to be acceptable.

Porous systems, which are adsorbent by physisorption, seem promising in terms of performance, provided the storage is done at 77 K, which is nevertheless a technological limit on a process that is otherwise practical and cheap. Notable progress remains to be seen with spillover nanocarbons, using aggregates based on palladium to absorb more than 8% at ambient temperature and under 80 bars (source NessHy (Novel efficient solid storage for Hydrogen FP7 EC Integrated Project) – 18/04/2008).

The compressed gas solution presents performance levels similar to the previous solution. It shows around a quarter (by mass) and a half (by volume) of the performance of the cryogenic solution. On the technological front, it is also the most economic solution in terms of primary energy. The inherent risks associated with any use of high pressure hydrogen, whether domestic or industrial, still have to be assessed.

It is difficult to establish reliable or pertinent economic criteria: the production and development costs, especially of systems, depend strongly on the quantities involved. However, certain types of "low-temperature" hydrides will constitute reasonably cheap means of mass storage.

It is interesting to note that the hydride modes can reasonably, technologically, and economically be realized. This highlights the often very variable performances of these different modes.

Not least of all remains the question of regeneration of energy (or saving primary energy), not only at the level of storage, which is only a link in the chain of "hydrogen as an energy vehicle", but also when developing generation-storage systems, storage-application systems, or even more integrated systems.

Figure 5.6. *143 years ago the horse drawn cart of Etienne Lenoir was the first vehicle in the world with an internal combustion engine that used hydrogen, tested near Paris in 1860*

5.8. Bibliography

Abundant literature covers some or all of the questions addressed in this short chapter. It is very difficult to present a complete list, especially as the opinions are sometimes contradictory, as for any new technology. It is possible to access important reports, which have copious information, via the internet.

[BUR] BURKE A., GARDINER M., *Hydrogen Storage Options: Technologies and Comparisons for Light-Duty Vehicle Applications*, http://repositories.cdlib.org/itsdavis/ UCD-ITS-RR05-01, University of California, Davis.

[RII] RIIS T., SANDROCK G., HIA HCG Storage paper (see International Energy Agency: http://www.iea.org/books).

[STU] STUBOS T., "Research on H_2 storage", in *The 6th Framework Programme*, NCSRD, Athens, Greece – see http://www.nesshy.net, http://www.storhy.net, with references to other European networks.

[TZI] TZIMA E., FILIOU C., PETEVES S.D., VEYRET J-B., *Hydrogen Storage: State of the Art and Future Perspective*, EC Joint Research Centre, EUR 20995 EN, http://ie.irc.cec.eu.int/ or http://www.irc.cec.eu.int/.

Chapter 6

Fuel Cells: Principles and Function

6.1. What is a cell or battery?

A fuel cell is a system that produces electricity and heat using a chemical energy source: the fuel. Cells, batteries, and accumulators are all very precise electrochemical entities, but they are considered to be more or less equivalent or have vague boundaries. The cell is an electrochemical device invented by Alessandro Volta in 1800 to produce electricity using a stack of electrodes and compartments (cells formed by the electrodes), which contains chemical reagents and which has now become commonly available in cylindrical or disk form. These containers enclose an initially fixed quantity of chemical reagents and can, therefore, only produce a limited quantity of electricity, up to the point where the chemical energy is exhausted. Only if this process, known as discharge of the cell, is reversible, by recharging the cell (which is done by injecting electrical energy by connecting it to an electric power supply) can the system again have the same capacity as it had initially. An electrochemical cell that is rechargeable is known as an *accumulator*. The term *battery*, which is borrowed from artillery, is used for both rechargeable and non-rechargeable systems. Theoretically, it refers to a set of cells that have been joined together, but it is also used for a single cell, which is why the different terms can seem to mean the same thing. Electrochemists distinguish between the systems with the help of a notion of order (which is unfortunately of little clarification): a "primary battery" is not rechargeable, whereas a "secondary battery" is rechargeable (accumulator). Note that in French, the word *pile* is reserved

---

Chapter written by Eric VIEIL.

for primary batteries, and the word *batterie* is reserved for accumulators, whereas in English there is no such distinction. Both primary and secondary batteries have the common characteristic of functioning with a non-renewable quantity of chemical reagents, which means that the former are no longer of use after their discharge, whereas the latter require a recharging phase, which will interrupt their production of electrical energy.

Chemical reagents that are continuously replaceable are known as *fuels* or *combustives* depending on their role in the reaction, as we shall see later on. Therefore, the fuel cell is a primary battery whose reagents are renewable during use. The primary/secondary distinction continues to be important for systems with renewable reagents, although these are rarely used for secondary batteries. A fuel cell functions in only one direction, that of converting chemical energy into electrical energy, whereas a *fuel accumulator* can function in both directions. Technologically speaking, this latter system is made up of two devices that can function independently: a fuel cell and an electrolyzer (the name given to a device that converts electrical energy to chemical energy). Table 6.1 lists the four categories that are founded in the criteria of rechargeability and renewability of reagents.

Process	Non-renewable fuel	Renewable fuel
Non-rechargeable	Primary battery	Fuel cell
Rechargeable	Accumulator (secondary battery)	Fuel cell + electrolyzer

Table 6.1. *Four categories of electrochemical systems for energy conversion*

Any storage process that is not purely electric is based on a double conversion of energy: in one direction in order to obtain a form of energy that can be stored, and in the other direction in order to put it back into the form of electricity. The fuel accumulator, i.e. linking a fuel cell with an electrolyzer, is therefore an electrical energy storage system in chemical form.

6.2. Chemical energy

The bonds between atoms forming a molecule, or the bonds between molecules forming another molecule, are the basis of the notion of chemical energy. In order to be able to bond, two atoms share a certain amount of energy, which is characteristic of the nature of the two atoms, and which we call the bond energy. When an atom changes partners to form a different molecule, the bond energy will change. In order to form a molecule that requires a greater bond energy than the initial molecule, it is necessary to supply energy to the system, and in the opposite case the system will release some of its energy. This is the basis of any chemical reaction, whether it is in the laboratory, in an industrial reactor, or in a biological cell.

As a substance that is capable of reacting chemically rarely consists of a single molecule, the expression for the quantity of chemical energy is split into two quantities, known as state variables of the system: the first is the amount of energy per molecule and the second is the number of molecules. The first is known as the chemical potential, μ, and the second is the number of moles, n (or N depending on the author). A mole is a number of molecules chosen conventionally to be equal to Avogadro's number, which is approximately 6.022×10^{23} molecules. So, the product of these two variables gives the total chemical energy, U, of a substance (U is the symbol for internal energy, also known as potential energy, of a system). In order to be rigorous and to respect the thermodynamic properties of the energy, it is convenient to express this product, not as a product between integral quantities, but between variations in some of these quantities, which is written:

$$\mathrm{d}U = \mu\, dn \qquad [6.1]$$

The chemical potential, μ, is therefore an amount of energy per mole, and its unit is the joule per mole (J mol^{-1}). It is a value that is linked to the nature of the substance but also depends on the number of molecules present and on the thermal agitation, i.e. the temperature. At least, for the ideal case of a substance that is not interacting with the environment, and whose molecules are not reacting between themselves. This is the case, notably, for a perfect gas, or for a solution that is sufficiently diluted for its components to behave independently. In the opposite case, the substance is real, and the chemical potential becomes dependent on other variables such as pressure, electric charge, surface energy, etc. A relation that is due to Gilbert Newton Lewis (1907) expresses these dependencies:

$$\mu = \mu^{\ominus} + RT\, ln\, a \qquad [6.2]$$

The chemical potential with a superscript of zero-barred is the *standard chemical potential* for the given substance, a value that only depends on the chemical nature of the substance, and R and T are the gas constant (around 8.32 J mol^{-1} K^{-1}) and the temperature, respectively. The argument of the natural log is the activity, a, a value that quantifies the influence of the number of molecules and their interactions. In the case of an ideal substance, the activity is simply proportional to the number of moles. As the proportionality constant depends on the choice of origin for standard chemical potentials, these are chosen to give a constant of mol^{-1}.

Let us take the simple case of a chemical reaction of transformation from a substance A into another substance B:

$$A \rightarrow B \qquad [6.3]$$

The variation in chemical energy that accompanies this reaction is written:

$$dU = \mu_A \, dn_A + \mu_B \, dn_B \qquad [6.4]$$

where energy is by definition additive, whatever its form.

Conservation of matter gives:

$$dU = (\mu_B - \mu_A) \, dn_B \qquad [6.5]$$

The value that will therefore determine the change in the quantity of chemical energy during a reaction for a given number of moles is the difference in chemical potential between the two substances. In order to predict the energies and to compare them to other processes, chemists use a value with the complicated name of "free molar enthalpy of reaction", defined as being this difference between chemical potentials (free enthalpy is also called Gibbs free energy). It is the total internal energy, combining thermal and hydrodynamic energies, which are not as readily liberated as chemical energy:

$$\Delta_r G_{AB} = \mu_B - \mu_A \qquad [6.6]$$

According to Lewis' relation, we can express this value, by assuming the two substances to be ideal, as:

$$\Delta_r G_{AB} = \Delta_r G^{\theta}{}_{AB} + RT \, ln \frac{n_B}{n_A} \qquad [6.7]$$

Every reaction is therefore characterized by a standard free molar enthalpy of reaction, $\Delta_r G^{\theta}{}_{AB}$, which enables the quantity of energy that will be released or consumed to be evaluated, at least in the case where the quantities of the substances are equal.

Let us take the example that concerns us in particular, namely the reaction that forms water from its constituents of hydrogen and oxygen, for which the standard free molar enthalpy of reaction is equal, in normal conditions, to:

$$H_2 + ½O_2 \rightarrow H_2O \quad \Delta_r G^{\theta}{}_{H2O} = -237 \text{ kJ/mol} \qquad [6.8]$$

The fact that this relative value for the formation of water is negative means that the bond energy of a molecule of water is less than the sum of the bond energies in its constituents before reaction. This expresses what we already know: that water is more stable than a mixture of hydrogen and oxygen in proportions of 2 to 1. This also tells us that we can eventually recover this energy by making water.

6.3. The unfolding of a reaction

Simply placing the reagents together in a container is not generally enough to lead to the reaction for which they are intended, even if the thermodynamic prediction is for a release of energy. The reason is that breaking the necessary bonds in the reagents, to separate the atoms of hydrogen in the molecule of di-hydrogen and the atoms of oxygen in di-oxygen, does not occur spontaneously and requires energy to be supplied in order for it to occur. We call this energy the *activation energy* of the reaction, denoted by $\Delta_r G^{\neq}$, and this can be just as important, or more so, as the energy of the reaction. The diagram in Figure 6.1 shows the evolution of energy per mole during the reaction[1].

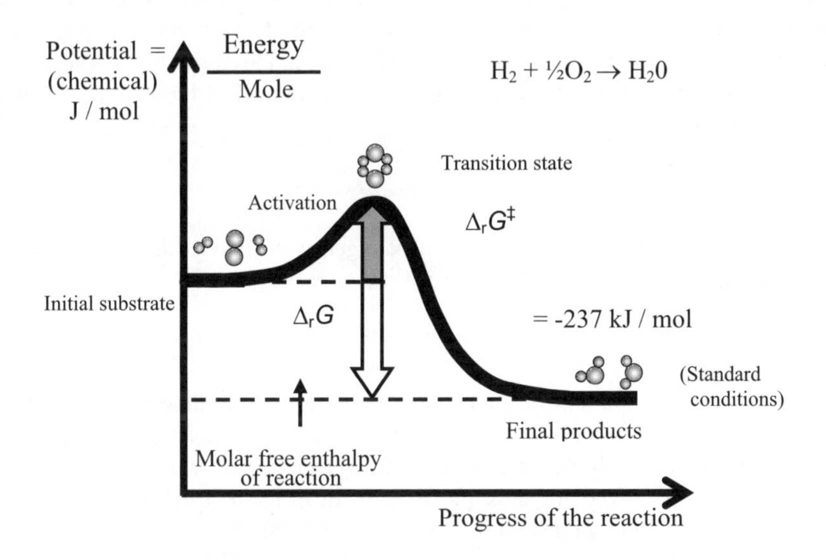

Figure 6.1. *Evolution of molar energy, the chemical potential, during the reaction to form water*

We note that the form of the energy profile leads to the nomenclature of an *activation barrier* that the reacting system must cross to proceed to the final product.

1 A frequent confusion, due to the influence of particle physics, is to make no distinction between an energy-per-mole (a chemical potential, a molar free enthalpy) and an energy. This would be analogous to confusing an electric voltage or a pressure with energy!

The activation energy may very well be supplied in a non-chemical form, for example, thermal, electric, mechanical etc., and the example of triggering the combination of hydrogen and oxygen with an electric spark is well-known.

Something else that is well-known is that this reaction produces an explosion, producing heat and hydrodynamic energy (increase in pressure or volume) from the very rapid combination once the activation energy has been supplied. This is the principle behind the thermal motor running on hydrogen.

The disadvantage of proceeding in this way is that the efficiency of a reaction, which is so rapid and badly controlled, is not at its optimum, due to its irreversibility[2]. In order to be able to control the reaction and to let it unfold in the best reversible conditions, we must take a look at the reaction details. We know that a bond between two atoms is formed by an exchange of electrons between the two. In fact, we note that a hydrogen atom, which possesses one orbital electron, may lend this electron to an oxygen atom, which needs two electrons to complete (make it up to eight electrons) its outermost electron shell. The act of a hydrogen atom becoming separated from its electron is written in the form of a reaction that produces two hydrogen ions, called protons, and two electrons, from one initial molecule of di-hydrogen:

$$H_2 \rightarrow 2H^+ + 2e^- \quad [6.9]$$

The acceptance of two electrons by an oxygen is written as the formation of an oxygen di-anion:

$$\tfrac{1}{2}O_2 + 2e^- \rightarrow O^{2-} \quad [6.10]$$

Therefore, the formation of water may be written in an alternative form to the reaction in [6.8], using these reactions (or rather, to use the electrochemistry term, these "half reactions"). It can be written as the association of two ions, meaning that two electrons are moved towards the oxygen in order to establish the chemical bonds of the final compound:

$$O^{2-} + 2H^+ \rightarrow H_2O \quad [6.11]$$

Any reaction that involves several atoms bonding can be decomposed into several elementary reactions, and there are other ways to write these reactions, but for the moment we will limit ourselves to the reactions we have just seen, as the principle of electrochemically controlling the reaction now arises. It is enough, for example, to recombine the last two reactions into a single reaction:

2 The efficiency of a hydrogen combustion engine is about 30%.

$$\frac{1}{2}O_2 + 2H^+ + 2e^- \rightarrow H_2O \qquad [6.12]$$

and comparing this with reaction [6.9] we understand that the global reaction for the formation of water may be controlled by the exchange of protons and electrons between the two initial reagents. Technically, the reagents need to be separated in two compartments to allow the exchange of ions and electrons between the two, in order to control the process and to prevent it from occuring too quickly. The diagram in Figure 6.2 shows the controlled combustion of hydrogen in principle.

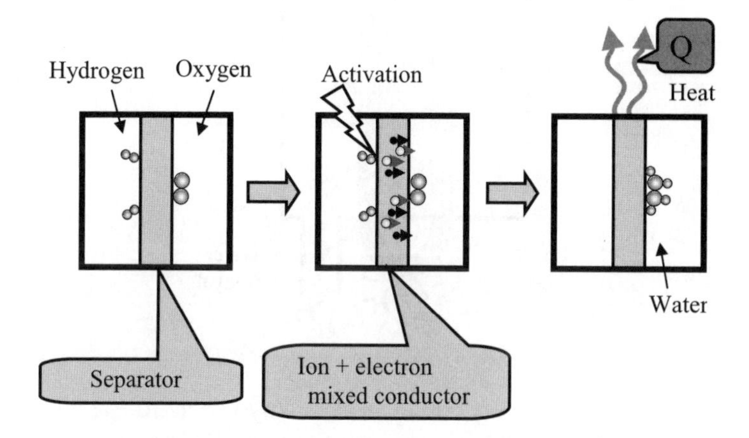

Figure 6.2. *Diagram of the three essential steps in the controlled combustion of hydrogen. Left: hydrogen and oxygen are separated in two compartments. Middle: the reaction is triggered by the supply of the activation energy, the protons and electrons cross the separator to combine with oxygen. Right: the compartment becomes occupied with the produced water and the heat produced escapes to the outside. There is no production of any other energy (hydrodynamic, mechanical, electrical, etc.)*

Clearly, the process depends on the correct functioning of the separator, which must be selective in only allowing the protons and electrons to cross it, but not the gases nor the water. In the kinetics of the process, we must consider the kinetics of the ionic transport, which is often the limiting step.

We have made progress in controlling the combustion, certainly improving its efficiency by using a reaction that is closer to being reversible but without recovering any energy other than thermally. However, it is sufficient to divert the electrons using an exterior electric circuit in order to extract the electrical energy. It is necessary for the separator to be the best possible electrical insulator and the best ionic conductor possible in order for this extraction to take place efficiently, and this efficiency can reach up to 60%. The addition of a circuit for the gases to be supplied to the compartments and of a route for the water formed to escape transforms the system in the previous figure into a fuel cell, which is said to be based on proton exchange.

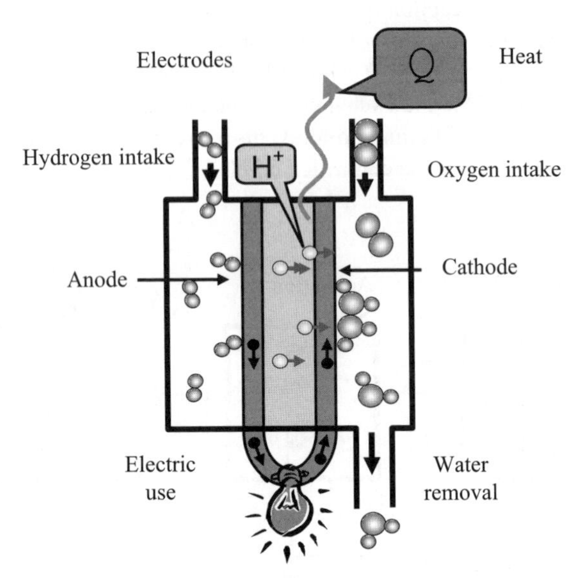

Figure 6.3. *Diagram of the principles behind a fuel cell based on proton exchange. The electrons are depicted in black and the protons are in white*

An electronic conductor, i.e. an electrode, is required in each compartment. The electrode that collects electrons is called the *anode*, whereas the electrode that reinjects them into the cell is called the *cathode*[3].

The variation in electrical energy is expressed identically to that of chemical energy (equation [6.5]) but with the corresponding state variables for the electrodynamic domain:

$$dU = V\,dQ \qquad [6.13]$$

The electrical voltage, V, is analogous to the difference in chemical potentials, and the electric charge, Q, is analogous to the amount of substance formed, n_B. Faraday's relation points to a constant of proportionality between these latter two

3 According to the terminology introduced by Faraday, from the Greek (the anode is the "way up" and the entrance for positive charges into the cell, and the cathode is the "way down" and the exit for these charges).

quantities, or their variations, which we present here, taking into account that for each oxygen atom two electrons are exchanged:

$$dQ = -2e\, N_A\, dn_B = -2F\, dn_B \quad [6.14]$$

The product of the elementary charge, e, and Avogadro's number, N_A, is known as the Faraday constant, and is roughly equal to 96,487 C/mol^{-1}. This lets us deduce the electric voltage that is available using the previous equations:

$$V = \frac{dU}{dQ} = -\frac{1}{2F}\frac{dU}{dn_B} = -\frac{\Delta_r G_{H_2O}}{2F} \quad [6.15]$$

The theoretical voltage under conditions of normal pressure (1 atm) and temperature (25°C), in an open circuit, and for equal amounts of the reagents is therefore given by:

$$V^\circ = -\frac{\Delta_r G^{\ominus}_{H_2O}}{2F} \approx 1.23 \text{ V (normal conditions)} \quad [6.16]$$

Under real conditions, in practice, the internal resistance of the cell can cause this voltage to drop by several tenths of a volt.

We have finished with the basic principles of a fuel cell, having shown how electrical energy can be produced using substances such as hydrogen and oxygen. The role of hydrogen is to give up its electron in order to make a proton and this reaction is known as oxidization. It could equally well have become known as an increasing reaction of the oxidation degree (i.e. the number of positive charges), but this term has not become common in usage. The inverse reaction is logically known as a reduction, which is what the oxygen atom undertakes by accepting electrons. A substance that is capable of giving up its electrons is known as a *reducing agent*, and one that is capable of accepting them is known as the *oxidizing agent*. These electrochemistry terms are sometimes substituted with the words *fuel* and *combustive*, respectively, and the latter qualification is attributed to any oxidizing agent that is capable of enabling the combustion of the fuel, which is, therefore, a reducing agent. It should be noted that the word "fuel" suggests that it is the primary material in an energy conversion process, whereas the reducing agent term is more general.

It is important to note that the reactions that we have described in the function of a hydrogen fuel cell can be reversed, i.e. that we can make a fuel cell function in the opposite direction in order to produce chemical energy using electrical energy. In the case of the reactions described, water and electrons are consumed to produce hydrogen and oxygen, a phenomenon known as electrolysis. The same device as that used for a fuel cell may be used as an electrolyzer, by applying, evidently, a larger voltage than that produced by the cell – at least 1.6 V. Linking an electrolyzer with a fuel cell that is linked to a reservoir of hydrogen (and eventually also of oxygen and water) can give an autonomous electrical energy storage system. The principle behind this "fuel accumulator" is illustrated in Figure 6.4 in the form of the *equivalent electric circuit*, where the conversions between electrical and chemical forms of energy are represented by "generalized transformers". The gas reservoir is represented by a "generalized capacitor" and "generalized switches" (gates) enable the separation of the phases of storage from the phases of use of electrical energy at exit. (It is naturally possible to allow usage occurring concurrently with storage.)

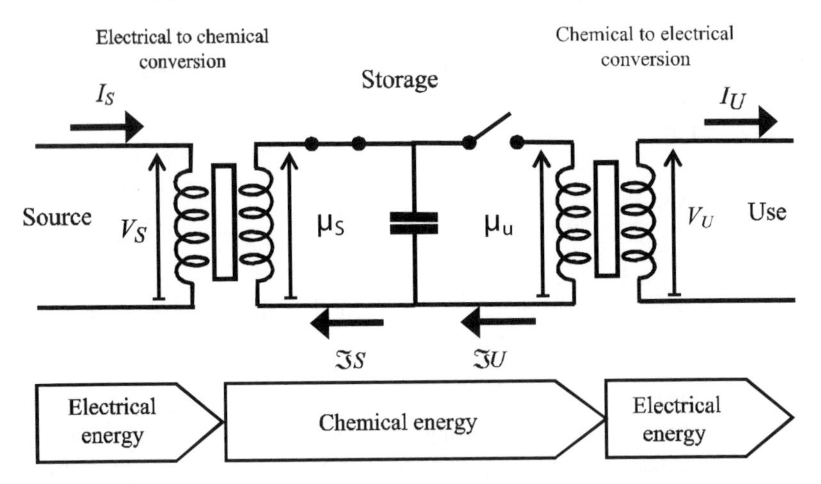

Figure 6.4. *Equivalent circuit for a conversion device that stores electrical energy in chemical form using an electrolyzer (left) and a fuel cell (right). The interrupter positions corresponds to the storage function*

The efficiency of the energy chain is evidently not excellent as, given the current state of the technology, the efficiency of an electrolyzer is inferior to that of a fuel cell (around 40-50%). For use as storage, it is a good idea to take into account the release of heat from the cell and the thermal requirements of the electrolyzer by coupling the two converters thermally. Moreover, heat that has been stored parallel to storage of the fuel can eventually be reused using thermoelectric conversion, and can therefore improve the global efficiency.

6.4. Proton-exchange membrane fuel cells (PEMFCs)

We have seen that the core process is based on the separation of the fuel from the combustive, and that the separator must be poor conductor for reagents and electrons and a good conductor for protons. The problem is that such a material does not exist, at least not with sufficient properties to work correctly within a fuel cell.

The demand for this extreme selectivity for some kinds of ions, while preventing other participants in the reaction from passing through, means that a very particular type of material is sought. It must have good mechanical performance in order to be able to resist the eventual differences in pressure, good temperature performance for those cells which require it, and must show sufficient resistance to the chemical aggressiveness of the reagents – if for the moment we ignore the economic cost and considerations regarding its life cycle – those are the main criteria for such a material.

The best separators with the best proton conduction are polymer membranes. The very best (and the most expensive, at €400/m²) is made using a product that is manufactured exclusively by the chemical firm Dupont de Nemours: Nafion. PEMFCs are manufactured by several companies, but at elevated prices and with a lifespan that is still too limited. (To give an idea of the orders of magnitude: the industry is hoping for the price to fall below €50/kWh and to provide more than 5,000 hours of function by 2010-2015.) The membranes used require a temperature of 50°C-80°C in order to allow sufficient proton conduction. A higher temperature would be even better, but we soon hit mechanical or chemical deficiencies at higher temperatures. With this type of fuel cell, it is possible to realize electric sources with low power (1 W) up to around a 100 kW. Principal applications that are targeted are onboard fuel cells in vehicles and for nomad applications.

Difficulties in making a separator working correctly with proton conduction have led to a search for other types of ion conductors that could ensure controlled combustion.

6.5. The solid oxide fuel cell (SOFC)

When we were detailing the electrochemical reactions that govern the formation of water, we had mentioned that the two partial reactions of oxidation of hydrogen [6.9] and of reduction of oxygen [6.10] and the reaction for water formation [6.11] that we repeat here:

$$H_2 \rightarrow 2H^+ + 2e^- \quad [6.17]$$

$$\frac{1}{2}O_2 + 2e^- \rightarrow O^{2-} \quad [6.18]$$

$$O^{2-} + 2H^+ \rightarrow H_2O \quad [6.19]$$

could be combined in several ways, and therefore that the reaction [6.12] on which a PEMFC is based was not the only way in which to control the combustion.

For example, combining reactions [6.17] and [6.19] instead of [6.18] and [6.19] gives:

$$H_2 + O^{2-} \rightarrow H_2O + 2e^- \quad [6.20]$$

This time it is the oxide ions, O^{2-}, which must cross the separator, from the cathode compartment to the anode, and the notable difference with the PEMFC is that water is formed in the anode (hydrogen) compartment. Such a cell, which relies on exchange of oxide ions, is named after the chemical nature of its separator as a SOFC.

Figure 6.5 schematically shows the principle behind an SOFC.

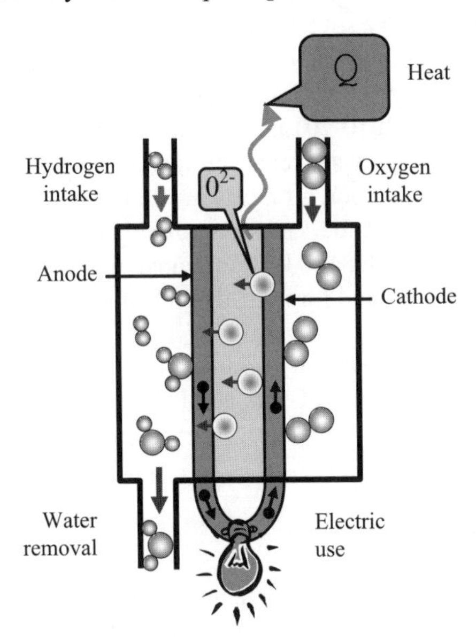

Figure 6.5. *Diagram of the principle behind a solid oxide fuel cell. Not only are the exchanged ions different, but also, unlike a PEMFC, water production takes place in the anode compartment, rather than the cathode compartment*

Materials that can conduct the oxide ion are themselves metal oxides (ceramic materials) and not only must they have the correct ionic conductivity, but they must also be sufficiently stable. Another important characteristic of these materials is that they must be put to work at high temperatures, within the range of 800°C-1,000°C.

A number of ceramic materials were proposed, and it would take too long to give details of all of these, as they are sometimes of very complex compositions, mixing several metal oxides and stabilized in a variety of phases. We list β alumina, cerine, strontium lanthanates, molybdenum oxides, vanadium oxides, yttrium-zirconium, mixed oxides etc. The main difference between them is their ionic conductivity, and above all their stability and their capacity to work at the lowest possible temperatures.

6.6. The alkaline fuel cell (AFC)

It may seem as though we should be finished with the possible combinations of one reaction with two others. However, we note that the reactions in question are not elementary reactions, as they all involve several partners. For example, instead of requiring the simultaneous addition of two protons, the reaction in [6.19] may be split into two steps as follows:

$$O^{2-} + H^+ \rightarrow OH^- \quad [6.21]$$

$$OH^- + H^+ \rightarrow H_2O \quad [6.22]$$

by using an intermediary, the hydroxide anion, OH^-. This ion can be exchanged between the two compartments provided that the oxygen compartment is supplied with water, and this leads to the following two half reactions:

$$H_2 + 2OH^- \rightarrow 2H_2O + 2e^- \quad [6.23]$$

$$½O_2 + H_2O + 2e^- \rightarrow 2OH^- \quad [6.24]$$

A fuel cell that uses this kind of exchange of the hydroxide anion is called an AFC. Its name originates from the need to function in an alkaline environment, which is generally realized by using potassium hydroxide, KOH, to increase the presence of the hydroxide anion in the anode compartment.

Figure 6.6 schematically shows the principle behind an AFC fuel cell. Compared to the other two types of fuel cells, we note the presence of a pipe at the bottom of the diagram, which is there to divert half of the water that is formed during the oxidation of hydrogen in order to send it into the cathode compartment in order to allow formation of the hydroxide ion, OH^-.

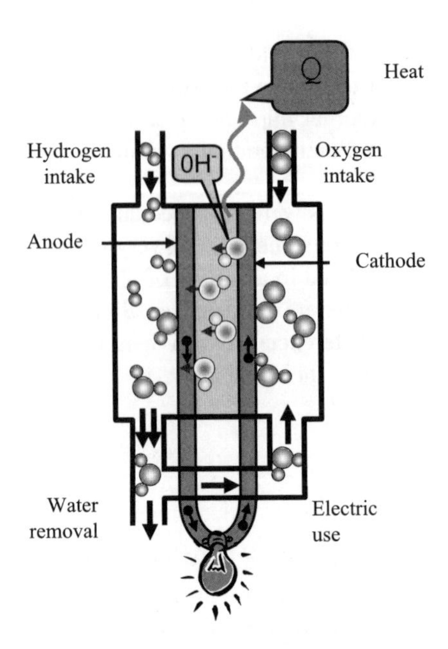

Figure 6.6. *Diagram of the principle behind a fuel cell, which uses exchange of hydroxide ions, otherwise known as an AFC. Note the recycling of half of the water produced in the anode compartment to supply the cathode compartment*

The separator in an AFC should be a good conductor of anions, and the materials that are capable of ensuring this function are principally basic polymers. However, unlike the proton membranes used in a PEMFC, the anionic membranes are more numerous and easier to make than Nafion. This gives a definite industrial advantage to AFCs, which should nevertheless be tempered by the fact that the alkaline surroundings are very corrosive for the metals in the electrodes and that the lifespan of an AFC is, therefore, limited.

6.7. Comparison of the different types of fuel cell

The diagrams in Figure 6.7 illustrate how the three types of fuel cell, that we have just covered, function.

These three types of fuel cell are not the only types, especially as the method of combining different reactions may be expanded by adding different reagents in another compartment. Without describing all the possibilities, Table 6.2 lists the three principal types that were already presented, as well as the *molten carbonate fuel cell*, (MCFC), and the *phosphoric acid fuel cell*, (PAFC). The application fields given are merely indicative of the most common use, but the fields are far from being exclusive.

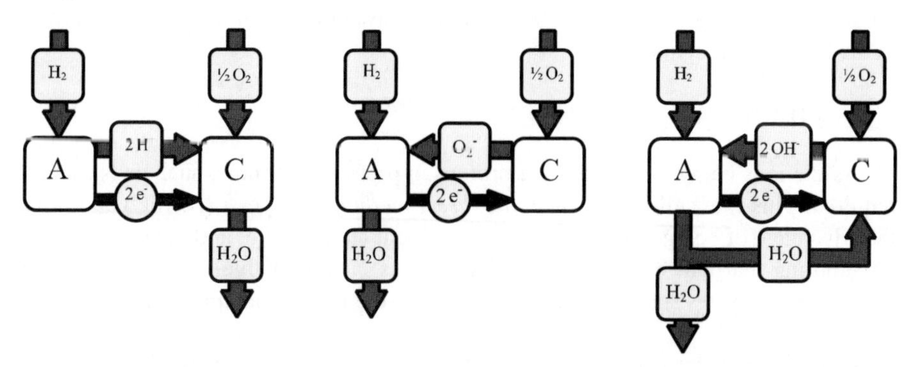

Figure 6.7. *The three main diagrams show the flux within the fuel cells. A and C denote the anode and cathode compartments, respectively: left: PEMFC; middle: SOFC; right: AFC*

Type of fuel cell	Temperature range	Efficiency	Electrolyte	Ion exchanged	Application
Proton-exchange membrane fuel cell (PEMFC)	50-80°C	50-60%	Proton exchange membrane	H^+	Nomad applications Electric vehicles
Solid oxide fuel cell (SOFC)	800-1,000°C	50-60%	Zirconium-yttrium	O_2^-	Electrochemical power stations
Alkaline fuel cell (AFC)	60-90°C	50-60%	35-50% KOH	OH^-	Space applications Onboard sources
Molten carbonate fuel cell (MCFC)	620-660°C	60-65%	Li_2CO_3/Na_2CO_3	CO_3^{2-}	Electrochemical power stations
Phosphoric acid fuel cell (PAFC)	160-220°C	55%	Concentrated phosphoric acid	H^+	Domestic/local electricity production

Table 6.2. *The main types of fuel cells and their main characteristics*

This table is limited to fuel cells that use hydrogen, but it is important to mention that other fuels are possible. Thus, the use of hydrocarbon compounds is perfectly possible, for example, methanol may be directly used instead of hydrogen, according to the following reaction:

$$CH_3OH + H_2O \rightarrow CO_2 + 6H^+ + 6e^- \quad [6.25]$$

As the ions being exchanged are protons, a proton membrane, such as Nafion, is required. This fuel cell is known as the *direct methanol fuel cell* (DMFC) and it has the advantage of being able to function at ambient temperature, and the disadvantage of emitting carbon dioxide, CO_2. Its use as a liquid fuel, which is easily rechargeable, makes such a cell particularly suitable for nomad applications.

Other fuels include methane, CH_4, which may equally well be used directly in a fuel cell provided that it is transformed into hydrogen, a process known as *reforming*, which is carried out at high temperature, within the temperature range of an SOFC. Coupling with that fuel cell gives an SOFC with *internal progressive reforming*. We will not list the other types of fuels as they are of little interest for storage, because their fuels do not allow reversible functioning of the cell.

6.8. Catalysis

We still have to cover an essential point that we have not touched on so far. We mentioned, at the start of our explanation of combustion, the need to overcome the *activation barrier* of the reaction by supplying extra energy. In a compartment of the cell, each reagent is isolated from the other and neither decomposition of hydrogen into a proton and an electron, nor decomposition of oxygen, come about spontaneously, except in very small quantities when the chemical equilibrium is displaced and the chemical potentials favor electron exchange. Creating these conditions is almost impossible when production of a little or some electricity is required. Therefore, it is essential to encourage these decomposition reactions by giving the oxido-reduction mechanisms the possibility to be carried out by other reaction paths, lowering the activation barrier and eventually accelerating the kinetics. The role of chemical compounds known as *catalysts* is to allow a chemical reaction to take place at a lower energy (per molecule) than that required in their absence.

The best catalysts for the oxidation of hydrogen are noble metals, such as palladium or platinum. These primary materials are very rare and expensive (with competition for them increasing practically endlessly), but they present the advantage of significant stability when faced with corrosion, compared with less noble metals. In principle, a good catalyst is not used up during the reactions that it facilitates; it is either released at the end of the reaction when it participated directly (complexation), or it serves as an adsorbent for the reagents to lower their chemical potential. As a result, very small amounts of the catalyst are necessary, so much so that it is often possible to alloy it with another less expensive compound.

The oxygen reduction reaction is also catalyzed by platinum, either alone or in combination with metal oxide supports or carbon. The reduction in size of each catalyst aggregate on the surface of an electrode, particularly on the nanometric scale, contributes to improvement of the efficiency and naturally reduces the amounts to be used.

A disadvantage of catalysts is that they are relatively fragile. They may be altered or made inert – we say they are poisoned – by the aggressiveness of the reagents or by the presence of impurities in the reagents. Carbon monoxide, CO, notably, but also sulfur compounds or some nitrogen oxides, are formidable poisons that can pose a number of problems when fossil fuels are used. This disadvantage is not present in the case of storage because the hydrogen produced by electrolysis is chemically pure.

The interesting thing about alkaline fuel cells is that they can accept catalysts that are less noble and, therefore, less expensive, such as silver, nickel, manganese oxides, and several others. As for platinum, they can be developed into nanometric aggregates on carbonated supports or on diverse oxide supports. Moreover, they possess the interesting property of not being as sensitive to poisons and impurities as other fuels.

6.9. Critical points

Fuel cell technology and their management are far from being completely mastered. Fuel cells are not yet truly commercialized and they are still the subject of for research. Let us begin by listing the critical factors that are of particular interest, by ranking them in order of decreasing difficulty (which depends on the type of cell, and therefore is not an absolute):

– efficiency, longevity, and cost of catalysts;

– good ionic conduction, mechanical solidity, and stability of separator;

– gas impermeability of the separator (SOFC);

– electric resistance of the separator;

– performance of the electric contacts (bipolar plaques);

– water-tightness of the separator;

– hydration of the membrane (PEMFC);

– heat management;

– water management and its escape.

Figure 6.8 shows how these points are located on a fuel cell.

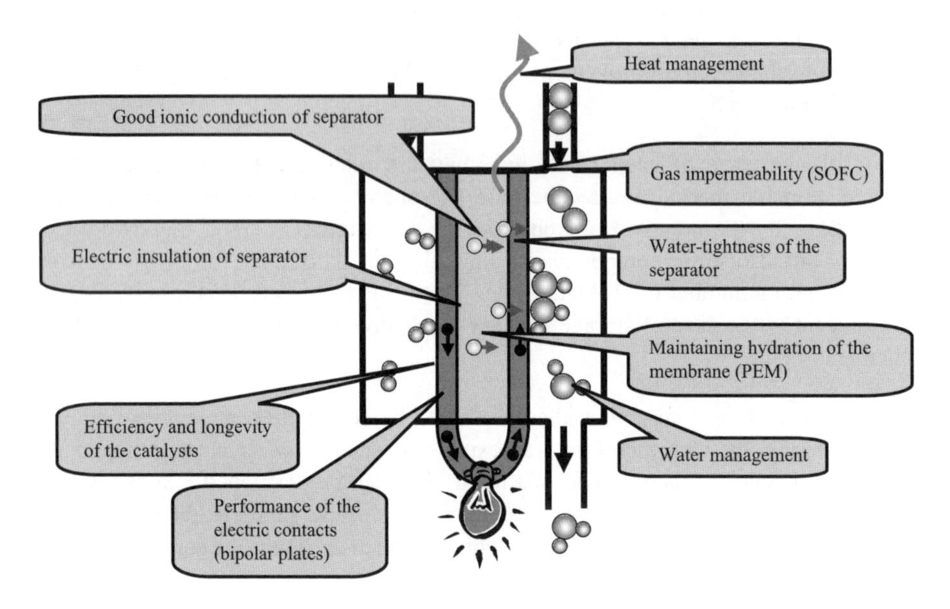

Figure 6.8. *The critical points in fuel cell technology (outside of cost issues)*

Among the important problems with a PEMFC is that of sufficient hydration of the membrane throughout the functioning of the cell. In order to ensure good conduction, the membrane must contain an amount of water. This varies depending on the temperature and the regime required of the cell. It must not fall below a certain threshold, because the less water there is, the less the conduction, the greater the debit of the cell and the greater the aqueous flux, which provokes rapid drying of the membrane and can lead to its destruction by perforation. Conversely, an excess of water in the pores or channels of the membrane will prevent the ionic sites on the membrane from playing their role in proton transport (drowning of the membrane).

Regarding SOFCs, the problems encountered relate above all to the high temperature and to the difficulty of making the ceramic materials, other than the membrane (electrodes, joints, and container), function at temperatures greater than 800°C. They also pertain especially to the stop/go phases of the cell, when there may be different expansions. For this reason, it is preferable to make a cell work slower rather than to stop it completely, which should not pose a problem in practice in a storage device.

Finally, alkaline fuel cells, which are of interest due to their much lower cost and their lower functioning temperature, unfortunately have the drawback of a shorter longevity than a PEMFC. The resistance to corrosion in an alkaline environment limits the lifespan of the electrodes and the composition of the environment must be closely controlled to prevent rapid degradation of the catalysts.

Regardless of the type of cell, the question of the thickness of the separating membrane is a compromise. Aiming for the best conductivity may not be correct, as what matters in the end is the conductance, G, of the membrane and not its conductivity, σ, even if they are proportional (for homogeneous membranes):

$$G = \sigma \frac{A}{\ell} \qquad [6.26]$$

A designer will be interested in increasing the area, A, and decreasing the thickness, ℓ, of the membrane in order to optimize the conductance. But the cost is also proportional to the area and the mechanical fragility is inversely proportional to the thickness! The best compromise between the thinnest thickness to ensure the greatest conductance and the thickest thickness to guarantee good mechanical performance is difficult to find and it depends on the applications and the types of function demanded.

6.10. Conclusion: the storage application

We have already mentioned a few points about fuel cells in relation to storage, and we recall these here briefly:

– linking an electrolyzer and a fuel cell forms a realistic device, the fuel accumulator;

– usage and storage may occur concurrently (unlike for a purely electrochemical accumulator);

– the heat released by the cell can be supplied to the electrolyzer (high temperature), which increases the overall efficiency (without this it would be around 25%);

– efficiency may be improved further by thermoelectric conversion of the residual heat.

We will now list a few characteristics that favor the fuel accumulator:

– storage does not have mass or volume constraints, nor are there restrictions on the mobility or the range, which exist for an onboard battery, and this leads to cheaper and easier realization;

– storage capacity is much higher than for an accumulator battery, for a comparable cost. It is only limited by the hydrogen reservoir;

– coupling with other sources of hydrogen production (bio-sources for example) gives flexibility;

– respect for the environment and durable development are facilitated.

The last point that we wish to discuss is the question of the regime under which a cell functions and its relation to the cell lifespan. The functioning regime of a cell, in terms of the power requested, but also in terms of the dynamics of the requests, influences the conductances and reduces them with use. The reason is that the chemical composition (presence of water, acid-base couples, annex ions) and physical composition (phases, grains, pores, joints) are influenced by the circulation of ions and their behavior is far from being linear (or predictable). This means that irregular requests, for example when used in a vehicle that has to stop and start frequently, are a source of precocious aging of cells, and that there is a long way to go before we know how to improve materials and how to understand the various phenomena. In this sense, the specifications for an electric vehicle are more of a narrow doorway than a help for rapid development of fuel cells. In contrast, storage is much less demanding an application, as we can more or less tailor the storage regime to the demands of the cell and we can even predict an acceleration in the development of cells so that hydrogen storage becomes the solution of the future.

Chapter 7

Fuel Cells: System Operation

7.1. Introduction: what is a fuel cell "system"?

Each of the five major types of fuel cell that are available on the market (see chapter 6) may be integrated into a fuel cell "system", which is dedicated to a particular market (or to a market sector) and which makes the most of that fuel cell's particular characteristics. Thus, fuel cells that work at low temperatures (AFC: *alkaline fuel cell* and PEFC: *polymer electrolyte fuel cell*), will be considered for both stationary applications and mobile applications in the transport domain, due to the fact that they can be started relatively quickly and that they respond well to thermal and electric cycles. In contrast, fuel cells that operate at much higher temperatures (MCFC: *molten carbonate fuel cell*, PAFC: *phosphoric acid fuel cell*, SOFC: *solid oxide fuel cell*) cannot handle rapid temperature rises and, therefore, require longer start-up times and are also more sensitive to thermal cycles. Therefore, their use is limited to stationary applications. Nevertheless, we must mention the particular uses that are planned for SOFC fuel cells: the solid nature of their electrolyte and the possibility of using carbon monoxide as fuel make these cells a candidate for transport applications, despite problems with heat management for these high-temperature cells (typically 800°C).

Chapter written by Daniel HISSEL, Denis CANDUSSO and Marie-Cécile PERA.

Fuel cells are intrinsically capable of supplying the demands for current with excellent dynamics. Unfortunately, it is not the same for the auxiliary functions that are necessary for them to work (hydrogen supply, air compressor, humidification system, cooling circuit, etc.). These auxiliary functions have different response times (from a few milliseconds to several minutes), which penalizes the "fuel cell system" function. So, we have come to the notion of a "fuel cell system", which integrates and includes the fuel cell itself, but also includes different subsystems (auxiliaries), which are required for it to work. The diagram in Figure 7.1 shows the different elements in such a system. We should underline the fact that by "fuel cell system" we generally mean a system such as that shown in Figure 7.1. This definition corresponds to the definition used by the FCTESTNET[1] project. Nevertheless, it differs somewhat from the definitions used by the IEC[2] or the SAE[3], proof, if it were needed, that much work must still be undertaken before competitive, high-performing, and well-defined fuel cell systems are seen on the market.

Referring to Figure 7.1, the fuel for the fuel cell must first of all be produced (especially if this is a hydrogen fuel – hydrogen being the most common compound on earth, but almost non-existent in its natural state) and stored. Then it is conditioned, in terms of pressure, temperature, rate of flow, and hygrometry, before being placed in the anode compartment of the fuel cell. The combustive must be conditioned in the same way, before being placed in the cathode compartment. Moreover, for each of the two gaseous circuits, we can recover the water produced within the heart of the cell by the electrochemical reaction, and carried by the exhaust gases of the fuel cell, in order to use this water to humidify the gases entering the cell. In certain cases and for certain modes of function, this leads to a system that is self-sufficient for water.

In addition, as the electrochemical reaction that takes place within the cell is exothermic, it is necessary to use a dedicated cooling circuit, with a liquid coolant, as soon as the cell's power becomes significant (typically above a kilowatt). Regulation of this cooling circuit aims to keep the temperature within the cell at roughly nominal conditions, as specified by the supplier of the cell. Of course, the cooling circuit controller must be coupled with the controller for the hygrometry of the entering gases, as the evolution of these two variables is, by their nature, coupled.

Finally, it is often indispensable to have a static converter to convert the energy from the fuel cell into electrical energy; the association of a fuel cell (energy source)

---

1 FCTESTNET: *Fuel Cell TEsting and STandardisation NETwork.*

2 IEC: *International Electrotechnical Commission.*

3 SAE: *Society of Automotive Engineers.*

with a power source (such as a supercapacitor or an Li-ion power battery) can be considered, depending on the target application.

Of course, the control and above all the management of the energy flux between all these subsystems and their interactions requires a dedicated supervisory device.

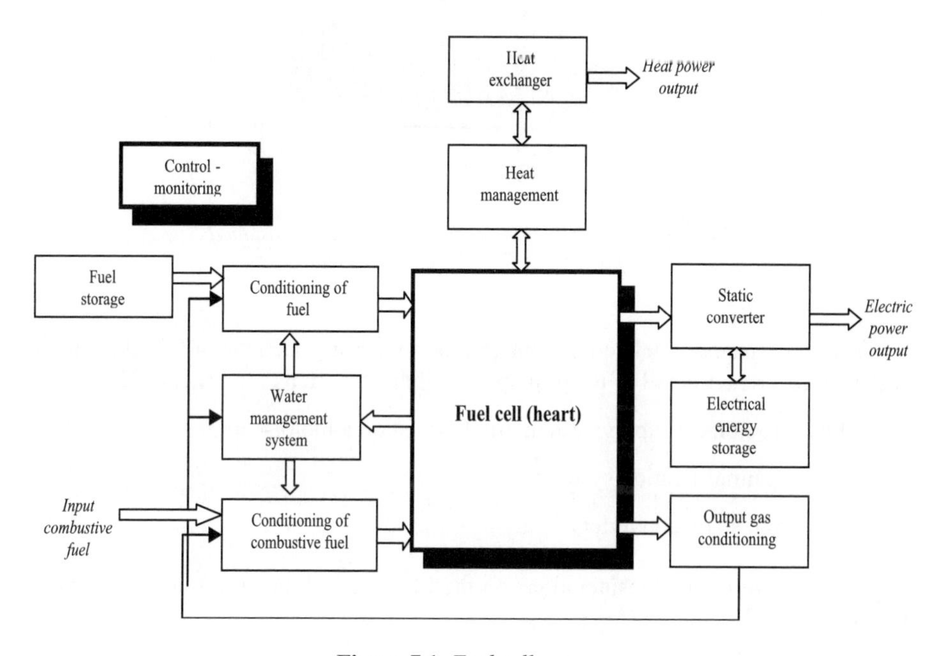

Figure 7.1. *Fuel cell system*

In order to be able to present ideas regarding the significant influence of all these auxiliaries on the behavior of a fuel cell, we can focus on the most common type of cell at the moment, especially if we are interested in transport applications: the PEFC fuel cell. Figure 7.2 shows the typical energy consumptions linked to the different auxiliary functions for this type of fuel cell system.

We first of all note that the net electric power only represents around two-thirds of the gross electric power, which is available at the exit of the fuel cell. Secondly, among the different auxiliary functions, the system for conditioning the oxidizer (which, in general, is based on an air compressor, for this type of application) is the greediest. Following that is the humidification system, the static converter, and the cooling system. Note that here we have assumed that hydrogen is directly available in compressed-gas form at the entrance of the system.

This illustration underlines the importance and the impact of the different auxiliary functions of the cell, on the dynamic behavior of the system and on the efficiency of the system.

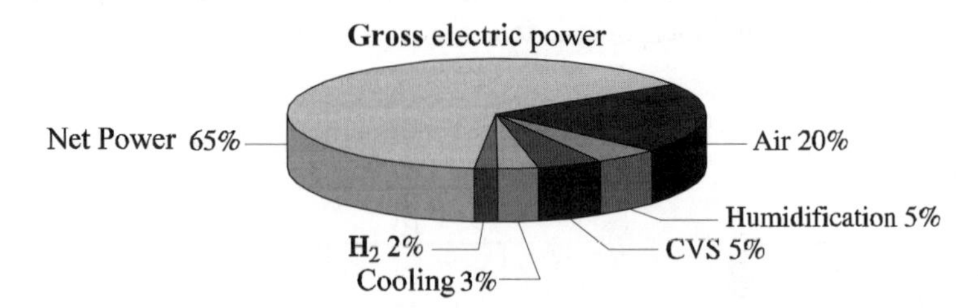

Figure 7.2. *Distribution of the produced and consumed power in a PEFC fuel cell system*

Having defined a "fuel cell system", in the remaining sections of this chapter we will provide greater detail of the principal auxiliary functions for a fuel cell system:

– the air (oxidizer) supply system, often based on a motor-compressor;

– the gas humidification system;

– the output static converter.

We will close with considerations on the lifespan and the failure mechanisms of the system.

7.2. Air supply system

7.2.1. *General considerations*

The first aim of this system (or circuit) is to supply an oxidizer to the cathode compartment of the cell. As for hydrogen, the oxidizer (pure oxygen or air) can be stored in high-pressure tanks. However, for transport applications (except for submarines or space applications), the oxidizer is often supplied from ambient air, which is abundant and free. Such a supply has a relatively low cost, but is still a delicate system. The main points to be underlined are as follows:

– Ambient air must be brought (in conditions that are compatible with the function of the fuel cell – we will come back to this later) to the reaction sites in the cathode compartment.

– Ambient air is often polluted with various gases or particles (CO, dust, etc.), which, if injected into the fuel cell, will reduce its performance and/or its lifespan.

– Ambient air must be "conditioned", in terms of hygrometry and temperature, in such a way as to improve the function of the cell (in certain cases, it is simply a case of purely allowing the cell to function).

Depending on their technology, their supplier, and their level of power, PEMFCs can function, in three modes:

– *"Respiration" mode*: as the name indicates, the fuel cell "breathes" ambient air according to its needs, linked to the demand for current. The supply of current provokes the consumption of the oxygen, which is present in the supply lines and creates a low pressure. This type of operation, which is very simple as it is totally passive and does not need any kind of air supply system, is reserved for cells with very low power (typically less than 200 W).

– *Atmospheric mode*: the air is supplied to the cell using air pumps or blowing pumps. The operation pressure for the fuel cell is very limited (a few hundreds of millibars at most).

– *Pressurized mode*: for cells with power greater than 1 kW, this mode is generally advised. It can be shown that the efficiency of electrochemical conversion is much greater when the partial pressures of the gases are high. Thus, this mode allows high power densities to be obtained for the cell, and, moreover, facilitates the hydro and thermal management of the cell (Figure 7.3) [NAS 00].

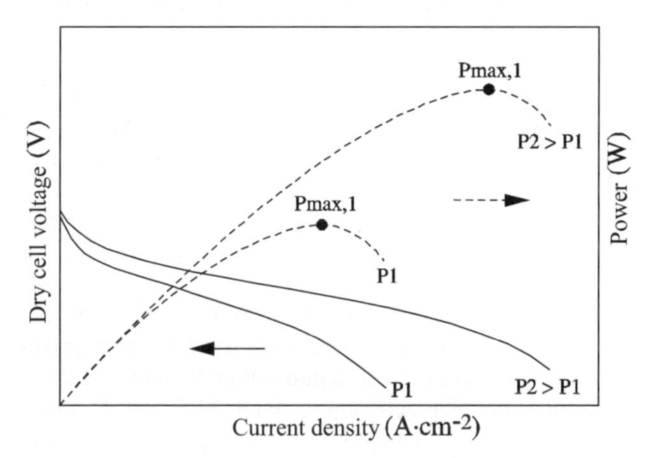

Figure 7.3. *Influence of pressure on the performance of a fuel cell [TEK 04]*

The compressed air is obtained using a motor-compressor. This allows the air to be compressed to different pressures (typically from 1.5 to 4 absolute bars) [WIA 00].

7.2.2. *Choice of a compressor that is adapted to fuel cell applications*

There are several technologies on the market for compressors, and the choice of air compressor is generally made depending on the air quality sought, the pressure, and the production capacity. The choice is made between the following technologies: centrifugal compressor, dry screw compressor, lubricated screw compressor, lobe compressor, pallet compressor, lubricated piston compressor, dry piston compressor, spiral/scroll compressor. The most widespread are volumetric compressors, in which the fluid is trapped in a closed volume that is progressively reduced in order to cause compression. In turbocompressors, the compression is obtained by converting the kinetic energy of the fluid into pressure using the speed of rotation of a wheel [DES 89].

The first factor in choosing a compressor for a fuel cell application is the quality of the air supplied at the exit. The electrochemical reactions that take place within the fuel cell are incompatible with having particles of oil in suspension in the air. In fact, residual droplets of oil in the compressed air may cover the catalytic sites and greatly reduce performance, and, eventually, the lifespan of the cell. Unfortunately, the majority of compressors that are "available off-the-shelf" are cooled and lubricated by oil. *De facto*, if we are looking for oil-free air, only non-lubricated solutions or those that are lubricated with water are suitable (such as turbocompressors, dry screw, dry lobe, dry piston, dry spiral, or membrane compressors).

Nevertheless, it seems obvious at this point that compressors lubricated with water could present one of the best solutions for fuel cell applications. Indeed, this technology has the advantage of conserving lubrication and therefore reducing mechanical losses, while giving quasi-isothermal compression. This enables the adiabatic efficiency to be increased and reduces the size of the humidifier, as the air will already be more or less humidified before entering the fuel cell [TEK 04].

Secondly, the quantity of air required and the range of the compressor power will help to determine the suitable technology. If usage is very intensive, rotating compressors and turbocompressors will require little maintenance, but the price will be high as a result. For fuel cell applications, high-speed compressors (from 10,000 rpm) such as centrifugal compressors or scroll or screw compressors are of interest [ZHA 03; DUB 06]. A regulator (or turbine) can then be mechanically (and eventually directly) coupled on the shaft of the motor-compressor and will recover some of the energy-pressure contained in the (low-oxygen) gases exiting from the cathode.

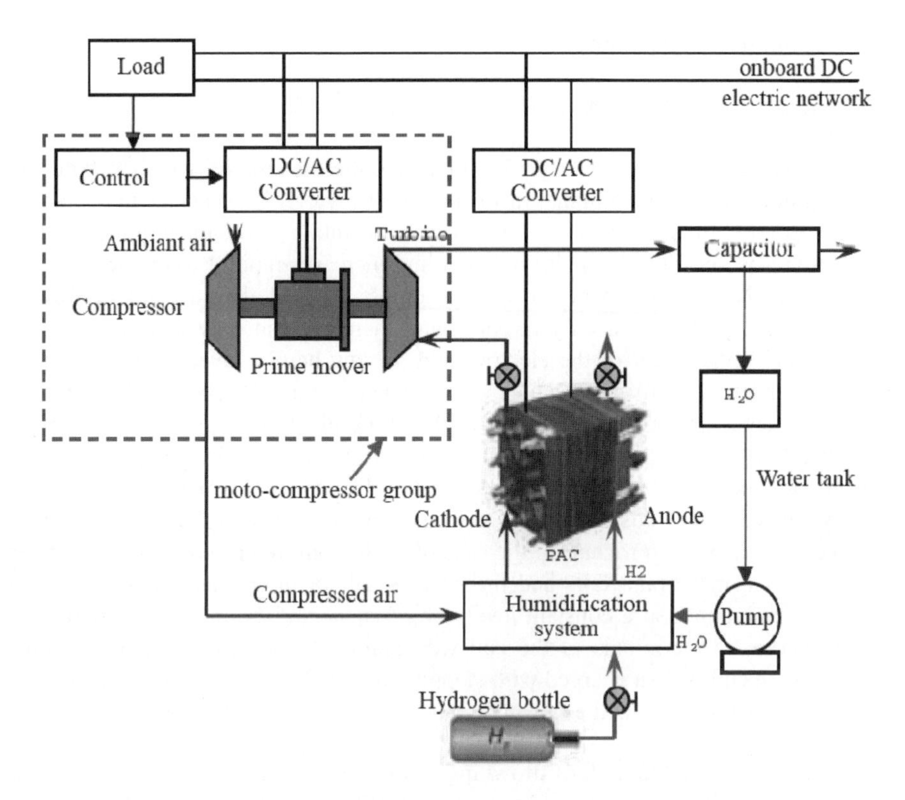

Figure 7.4. *Diagram of the principles behind a PEMFC using a compressor linked to an expansion turbine that regulates the air entering [TEK 04]*

Among the rotating compressors, the spiral, screw, and membrane compressors seem to be the better adapted to fuel cell systems in terms of efficiency (for low and medium power fuel cells) than piston or centrifugal compressors. These compressors are airtight, which means that their volumetric efficiencies are higher than other rotating compressors (lobe, pallet).

Finally, and above all, they are sized for ranges of powers that are suited to fuel cell applications. At the moment, there are a few prototypes of compressors that were specially designed for small fuel cell generators. In general, these compressors all have one or several bad characteristics (mediocre efficiency, excessively high noise level, lifespan that is incompatible with a fuel cell application, etc.). Therefore, there is still a lot of work to be done in this field!

7.3. Gas humidification system

7.3.1. *General considerations*

Continuing our study of the PEFC fuel cell, the electrolyte, which is solid, is made up of a polymer membrane. In such an electrolyte, it is necessary to guarantee good hydration of the membrane in order to maintain good ionic transfer. The conductivity is very low for a dry polymer and increases rapidly when the polymer is hydrated (Nafion polymers can absorb almost 20% of their weight in water). Moreover, the protons can only migrate (and therefore can only contribute to the current) under the effect of the electric field, if they have a layer of water around them (around two to five molecules of water per proton) [MOR 03]. However, a balance is necessary, as there is a risk of drowning or of drying out the fuel cell if the quantity of water is not correctly controlled. In the case of drowning, the access for the reagent gases to the catalytic sites may be blocked, which, when there is demand for current, leads to a fall in the potential at the level of the cell, which is relatively rapid and significant. In the case of drying out, the proton conductivity of the electrolyte membrane falls and the protons cannot migrate. If the demand for current is maintained at a constant level, the resistance of the membrane increases, the voltage at the terminals of the cell will again fall. In both cases, the control system of the cell is then charged with stopping it from functioning on detection of a voltage level (of the fuel cell as a whole, or of a constituent cell) that is too low.

From where does the risk of drowning or drying out originate? It comes from a combination of phenomena (see Figure 7.5 [HER 06]), which are directly linked to interactions between what is happening at the heart of the cell and the operation of the fuel cell system:

– Water forms at the cathode during the electrochemical reaction, in a quantity that is related to the current being supplied by the fuel cell.

– During the migration of protons from the anode to the cathode, there is also a transfer of water. This phenomenon is known as "electro-osmosis".

– Water diffuses from the cathode towards the anode due to the difference in concentrations between the two compartments.

– In order to prevent drying out of the electrolyte membrane, especially for high temperature operation, water is generally introduced into the reagent gases (at the cathode, at the anode, or even at both sides simultaneously), which unfortunately contributes to displace the equilibrium.

– The temperature operation of the cell and the temperatures of the reagent gases at their entry to the cell, equally contribute to displacing the equilibrium point.

Localized phenomena of evaporation/condensation take place at the heart of the cell, depending on the local temperature and pressure conditions.

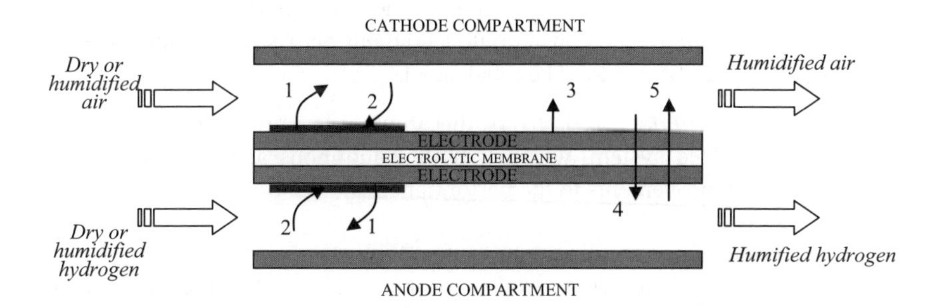

Figure 7.5. *PEMFC: water cycle in the heart of the cell. 1-2: evaporation-condensation of water on the walls; 3: production of water from the cathode side; 4: diffusion; 5: electro-osmosis*

7.3.2. *Possible humidification modes*

The humidification device is the second major component in the gas supply line for a polymer membrane cell and is equally the second consumer of gross power in the cell, in the case of an autonomous system of more than 1 kW. As for the air supply system, several architectures may be used: passive solutions that do not put a strain on the net power of the cell but offer less optimal control of the operation conditions compared to active solutions that are more efficient but more energy-greedy. We can classify several solutions, as follows:

– *Self-humidification* by the production of water during the reaction. Beyond a few hundred watts, it becomes hazardous to rely only on the internal production of water by the cell to maintain sufficient humidification of the membrane, unless a device is used to recover the water. Indeed, even in the case where the quantity of water produced is sufficient *a priori*, some of this water will escape the cell.

– *Humidification of the fuel:* in the majority of applications, it is necessary to supply water to the cell, using external humidification of the gases entering the cell. As water is produced at the cathode, the zone that is initially lacking in water and the area of the membrane that is susceptible to drying out is the anode, which is why humidification of the fuel is necessitated. First of all, in order to control the risk, which is linked to the presence of hydrogen, this supply line must be simplified as much as possible by limiting the number of components that could become the origin of leaks. Conversely, in the case of excess water, the membrane is no longer going to absorb water. Therefore, the excess must be evacuated via the anode circuit, which is rarely in open mode but more frequently in closed or recycling mode, in order to limit the consumption of hydrogen. Therefore, evacuation by the anode proves to be more complicated.

– *Humidification of the oxidizer*: this is the most frequently used solution. Diffusion of water due to the concentration gradient across the membrane enables the deficit of water at the anode to be rebalanced.

– *Humidification of the two reagents*: this solution is more rarely used because it leads to a complicated system, which is both voluminous and expensive, without bringing definite improvements to the performance.

We can classify the humidification systems into two categories, depending on whether they require direct water exchange or a buffer water reserve. Membrane exchangers and enthalpy wheels belong to the first category and, assuming that the cell is self-sufficient in water, systems for injecting water and distillers belong to the second category.

7.3.3. *Membrane exchangers and enthalpy wheels*

The principle behind these systems is to transfer the water contained in the oxidizer downstream from the cell, which is low in oxygen but rich in water vapor, towards the dry oxidizer entering into the cell. Certain devices are completely passive and are made of porous membranes [HUI 08]: the humid flow passes on one side of the membrane, the dry flow passes on the other side, and the membrane enables the transfer of water due to the concentration gradient from the humid flow to the dry flow. This solution was used for humidification of the air in the autonomous Nexa™ power module of 1.2 kW (Figure 7.6).

Figure 7.6. *Autonomous Nexa™ power module of 1.2 kW, using a membrane exchanger in the air circuit*

The enthalpy wheel, as its name indicates, consists of a rotating element that improves exchange. It is made up of a porous cylinder, covered in a desiccant material, which turns slowly inside a casing. The hot and humid air coming from the cell penetrates into the wheel. The drum absorbs the water vapor contained in the exhaust gas and transfers it into the entering dry airflow. The other components of the exhaust gas are evacuated, as a result of the presence of the covering. Certain models include regulation of the speed of the wheel, thus providing a control variable for the quantity of water exchanged.

These devices simplify the system and reduce the number of components compared with systems that need a water reservoir. There is no condenser, they consume little, if any, energy, and recover some of the enthalpy from the exhaust gases. However, they do not allow the dew point to be controlled or the temperature of the entering gas; these parameters depend on the temperature and the level of humidity of the exhaust gas, i.e. on the operating point of the cell [GLI 05], [STU 08]. This does not pose a problem when the cell works at an operating point, when the variations in the loads are slow or else at very high frequency. Conversely, when the current is fixed at some level, or when the medium frequency solicitations are of the same order of magnitude as the time constants for water diffusion across the porous materials used, this coupling can introduce instability in the regulation of the system.

7.3.4. *Systems with a reservoir*

Systems that use a water reservoir offer the possibility of fixing the dew point of the gas entering into the cell, independently of the operating point of the cell, both with regard to temperature and with regard to the current inducing the quantity of water produced. We can distinguish two systems: water injection and distillers.

In a system using water injection, regulated air enters at the base of a column while water vapor is injected at the top. As the air rises in the column, it becomes humidified [JUN 07]. Water vapor exchange is favored by the presence of metallic rollers in the column. The amount of water to be injected is calculated depending on the temperature, pressure, air flow and the humidity level sought. A separator, situated at the base of the column, recovers the water that is not transferred. The temperature of the jet of vapor is one of the control parameters of the system. The humidity level of the gas entering the cell is not known precisely. Indeed, the presence of water in the separator shows that some of the water is not transferred and that there is a gap that is difficult to measure between the real humidity level and the fixed set point.

Another type of system consists of using a distiller. This solution enables the temperature and the humidity of the gas to be fixed simultaneously and independently as well as minimizing the duration of transitories (Figure 7.7). The delivered dry air is fixed by a delivery regulator. The air penetrates into the distiller where it is saturated with water at the water temperature of the distiller, which is the highest on the circuit. Then it is cooled in an exchanger, which thus fixes the dew point temperature of the gas. The liquid water is evacuated into a separator. Finally, a reheater corrects the temperature at the entrance of the cell. This system is regulated by two easily measured parameters: the temperature at the exit of the condenser and the temperature at the exit of the reheater (circled in Figure 7.7). The fact that only these two temperatures are controlled, leads to the qualities of reliability and rapidity that are being sought. This solution thoroughly controls the operating conditions and their repetitivity. Nevertheless, it is particularly voluminous and energy greedy because it requires a significant volume of water to be kept at a particular temperature, and because of the number of components. It will therefore be reserved for experimenting in the laboratory [MOR 03; GLI 05].

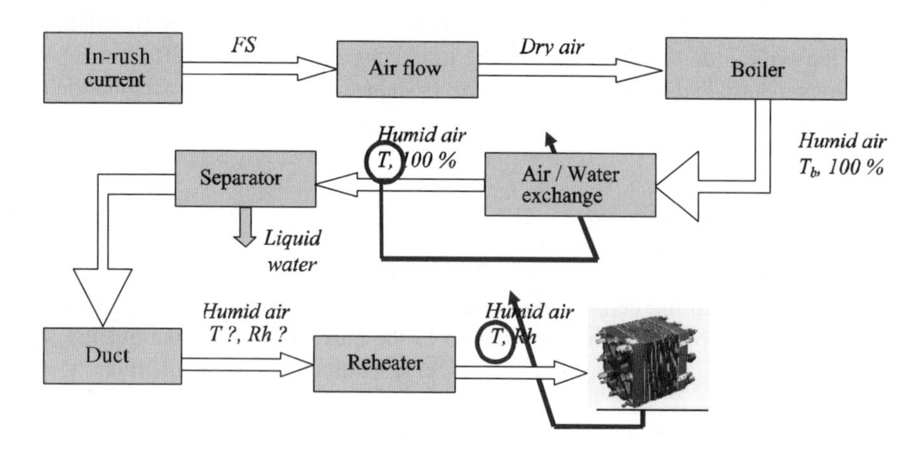

Figure 7.7. *Principles behind the operation of a distiller humidification system permitting the dew point and the temperature of the airflow to be fixed*

7.4. The static converter at the stack terminals

The fuel cell is characterized from an electrical point of view by a voltage level that is relatively low and significant current levels. These levels are estimated to be around 0.7 V per cell and 1 A/cm^2, depending, of course, on the fuel cell used.

Therefore, it is necessary to define particular static converters, meeting these demands, which present reduced volumes and weights and increased efficiencies. As a chapter of this book is devoted to this area, we will not revisit it in this chapter.

Moreover, solutions that authorize operation in degraded mode, or even a continuity of service at the level of the fuel cell, must be conceived. Linking fuel cells via adapted power electronics can resolve this and offer redundancy in the global system [CAN 08].

7.5. Lifespan, reliability and diagnosis

Progress on fuel cell systems between the 1990s and today, has enabled their use in new fields, such as mobile computing, stationary cogenerators, and vehicles. However, these remain prototypes and their large-scale industrial viability is prevented by one of the major problems that exists for fuel cell technology: their lifespan. This is notably subject to the conditions of use of the cell and the size of its stacking. Thus, to be of interest in onboard applications, fuel cells must attain a lifespan of at least 5,000 hours for a car and of 20,000 to 40,000 hours for public transport. Today, even if mono-cells with weak active surfaces can already satisfy these criteria, the stacks of PEM cells can only hope for a shorter lifespan, of the order of a thousand hours [BON 08; ESC 05; WAH 08].

Research is still required to improve the performances of cells in terms of lifespan and reliability, and to reproduce the conditions of their use in testbeds (with, for example, in the case of the vehicle application, the replication of the dynamics of the load current linked to the time-speed profiles for missions of the vehicle [WAH 08], the study of the impact of vibrations on the mechanical behavior of the cell [ROU 08], the problem of operation and cold start [BÉG 08]).

From a systems point of view, the main objective is then to research and ensure optimal operating conditions, leading to higher efficiencies at the system level and to increased reliability and longevity. From this perspective, the technological choices concerning the different necessary auxiliary functions (devices for conditioning the reagents, electronic converters, energy storage elements, etc.) must be evaluated and adapted in order to locate the cell in the best possible operational environment. Different faults may appear at the heart of the cell itself, as a result of the constraints being applied, but they may equally appear in peripheral devices. These problems require the development of experimental diagnostic methods that will allow the faults and breakdowns to be qualified. Moreover, more global diagnostic, and even forecast strategies must be developed.

7.5.1. *Faults and their origins*

Faults that appear at the level of the stack can be classified in different ways, by considering the physical nature of the phenomena leading to a degradation (mechanical, thermal, electrochemical, etc.) or by their degree of gravity (reversible degradation in performance disrupting the performance stability, or a degradation that leads to irreversible lowering of performance, or else: impossibility of operation inducing reliability problems in the generator). The speed at which degradation spreads can also be a criteria for classification of different faults [LAC 03], [WIL 03]. At the level of the stack, we will only consider the most commonly encountered faults from now on.

As already mentioned, water management in the PEM cells is a problem that is both complicated and crucial. It has major consequences in terms of obtaining both high and stable performances. Drowning is generally localized in several cells and impedes the access of reagent gases to the interfaces where the chemical reactions take place. It then leads to fluctuations and dispersions of the cell voltages, often suddenly and unpredictably, and basically amounts to an instability of performance of the cell especially at high current density. Drowning can generally be reversed and the initial performance can be attained again by acting on the system: by reducing the humidification of the reagents (either by using the temperature gap in the cell and in the gas humidifier(s), or by using purges with the reagents in excess for a few seconds), by changing the way the pressure gradient between the anode and cathode is managed. However, repeated drownings can lead to operating at sub-stochiometric conditions. Even when localized, medium-term or long-term drownings can have major consequences on the degradation of the constituent elements within a cell.

By contrast, insufficient humidification of the reagents or excessive cell temperatures can lead to drying out of the membranes and reduce their conductivities. Provided these drying out phases are not too long, and are not recurrent, the performances of the cells can normally be re-established by restoring the correct hydration of the membranes. Nevertheless, cycles in drying out/hydration are particularly difficult operating conditions. They reduce the lifespan of the stacks due to mechanical stresses and due to the formation of hot points at the centers of the cells. The system parameters that enable the water load at the heart of the cell to be controlled are the same as those given in the case of drowning.

The quality of the reagents in terms of purity with respect to contaminants, such as carbon monoxide, is another important factor to take into account in order to improve the lifespan and performance (they are tolerated up to a maximum of around 10 to 50 ppm of CO). The presence of too much contaminant leads to reduction in the electro-active surfaces in the membrane-electrode assemblies. This lowers the catalytic activity, but the effect can be curbed by application of adapted purging mechanisms.

If faults occur within a PEMFC fuel cell, they often have a link to the system, its architecture, and its control. Thus, a poor control of the pressure gradients between anode and cathode can mechanically damage the membranes, which are fragile systems due to their thickness (25-100 µm). Badly adjusted delivery of gaseous reagents leads to substochiometric conditions. Operating with transitories at small current densities, or even no current at all (at OCV – *open current voltage*) has an impact on the lifespan. Mastering physical parameters is particularly delicate when dynamic cycles (strong variation in current, flows, temperature and pressure gradients) are applied to the fuel cell system.

A cell under nominal and stable operating conditions still suffers from the effects of aging at the component level, i.e. diffusion layers, electrodes, membranes, joints. Endurance tests show that there is a reduction in the properties linked to the transportation of reactive elements from the bipolar plaques towards the reaction interfaces. There are reductions in the catalytic activity during operation, there is corrosion of the metallic bipolar plaques, and the impermeability between the anode-cathode-cooling circuit compartments is compromised.

At this point, it is also important to mention that in present fuel cell system prototypes, auxiliaries can also represent sources of dysfunction. In practice, untimely stoppages and periods where use is impossible are not only due to faults in the heart of the cell…far from it!

The impact of physical parameters on the lowering of cell performance, on the appearance of faults, on the aging of the cell and of the system, must be studied using a variety of experimental methods for characterization, using methodologies in the organization of tests [WAH 06] and using adapted diagnostics [HER 06].

7.5.2. *Experimental methods for characterization*

The experimental methods of characterization that are most commonly used are based on polarization curves, impedance spectrometry, and study of the response of the cell to dynamic solicitations and to voltammetry [WU 08].

The polarization curve (charge current-voltage of cell) gives a global image of the static performance of a fuel cell, but it does not allow the different voltage drops to be clearly dissociated. These voltage drops are due to different types of losses, mainly: activation phenomena, permeation of reagents across the membrane, Ohm's law, diffusion of reactive elements into the heart of the cell. Therefore, other methods of characterization and of diagnosis may be used to better determine the effects of different physical phenomena on the voltage of the cell.

Impedance spectroscopy is generally used in order to estimate the state of hydration of the membranes, to detect eventual difficulties linked to the transport of reagents into the cell. Impedance spectroscopy is a measurement technique which consists of imposing a periodic input value (for example a current) on a stable system (here, the cell) and analyzing the corresponding output value (galvanostatic voltage). The ranges of frequencies explored for characterization of fuel cells vary typically from 30 kHz to a few millihertz. The resistance of the membranes is determined at frequencies that are generally between 1 and 10 kHz, whereas charge (electrons, protons) transfer phenomena and diffusion phenomena take over at lower frequencies.

Impedance spectroscopy is a technique that has been adapted to the detailed study of degradation mechanisms. This method is largely used by the electrochemist community to study electrochemical generators such as batteries, accumulators, or fuel cells. In the latter case, the studies are often undertaken on small assemblies with small active membrane surfaces (of the order of a few centimeters squared), or even on demi-elements (anode or cathode). A much smaller number of studies concern stacks with several cells. The use of impedance spectroscopy on stacks is particularly interesting as the measured impedance corresponds to an average across the cells, due to the different inhomogeneities that exist within the stack (temperature, fluid distribution, etc.). Moreover, taking into account the present fabrication processes, the cells may have non-negligible disparities between them in terms of performance. The study of the response of the cell to dynamic solicitations for square wave currents is an alternative approach that allows different causes to be identified for the voltage surges [MOÇ 07].

Voltammetry is another technique that is used in the laboratory in order to characterize fuel cells. For example, it allows the active surfaces of an electrode to be determined. It is also used to evaluate the permeation current linked to the reagents moving across the membrane (*crossover*).

7.5.3. *Diagnostic methodologies and strategies*

Benchmarking of experimental procedures that allow the detailed characterization of fuel cells is required, but a definition of diagnostic tools aiming for applications is also necessary. Some of the detailed characterization methods may be adapted or partly used to monitor and control stationary or onboard generators. From this perspective, *ad hoc* experimental procedures, which are efficient and simplified, need to be defined to allow the localization of faults in the heart of the cells, the determination of causes for the fault, and the path to be taken (stopping the device, eventually starting an operation to recover performance and/or operating in degraded mode). The interest of the impedance spectroscopy technique is clear in the laboratory, but its use on real systems, onboard for example, is more difficult. Some current solicitations, which are linked to electronic converters that allow the conditioning and use of the electrical energy issued from the fuel cells, may be used to measure the impedance, especially of the membranes, or even to control the humidification systems and to keep the water load adequate in the electrolytes [SCH 05]. Another example, the application of transitory reagent flows or of small variations in the anode-cathode pressure, coupled with monitoring of the no-load voltage of the cells, allows faulty cells within the stack to be detected [TIA 08].

Today, the diagnostic methods used on prototype fuel cell systems depend partly on monitoring of different parameters judged to be important (for example, temperature of the cell, pressures of gases at the entrance to the stack, etc.) and also depend partly on observation of the voltage response of individual cells in order to evaluate, in real time, their operating state (*state-of-health*). Ranges of operation are defined for each of the parameters and control of the generator is realized using different critical thresholds (for example, minimum voltage threshold for the voltage levels in the stack). A reduction in the voltage of a cell, for example, allows identification of eventual poisoning of the cell by CO, eventual dehydration of the membranes, or drowning of the cell, etc. The control system is then called on to compensate for these problems by acting on the different auxiliary functions (sub-system for processing the fuel, cooling circuit and/or humidifiers, etc.) and by ordering different procedures (reduction in the load current intensity, triggering an air purge, etc.). The difficulty linked to such controls resides in the fact that the information collected is not always discriminating as to the origin of the fault, and that it does not always allow an anomaly to be anticipated.

In order to deal with this problem, one possible way is to implant more sensors within the fuel cell system so as to better detect any gap between the conditions and normal operation. However, such an approach leads to problems at the design level of the cell and of the entire system (more expensive generator, but also more complicated generator, which is therefore potentially less reliable *a priori*). Therefore, other diagnostic strategies must be considered. One solution consists of defining physical or behavioral models of the system (using a series of static and dynamic experimental trials, in both normal and degraded operating mode), putting these models in a computer, and using them in real time to compare the output of these models (cell voltages for example) with the values measured in the system in operation. The use of a physical model that describes the phenomena with mathematical models leads to an interest in understanding the explicit causality within even the model of the system. The diagnostic process is often simplified but, in return, the modeling is complicated [HER 06]. A behavioral model (black box) is without a doubt easier to create as it is linked more directly to the experimental trials carried out on the system [FOU 06; HIS 07]. However, in that case, the absence of an explicit causality makes localization of faults a more difficult step to undertake.

7.6. Bibliography

[BÉG 08] BÉGOT S., HAREL F., KAUFFMANN J.-M., "Design and validation of a 2 kW-fuel cell test bench for subfreezing studies", *Fuel Cells From Fundamentals to Systems*, vol. 8, no. 1, pp. 23-32, 2008.

[BON 08] BONNET C. *et al.*, "Design of an 80 kWe PEM fuel cell system: scale up effect investigation", *Journal of Power Sources*, vol. 182, issue 2, pp. 441-448, 2008.

[CAN 08] CANDUSSO D. *et al.*, "Fuel cell operation under degraded working modes and study of a diode by-pass circuit dedicated to multi-stack association", *Energy Conversion and Management*, vol. 49, no. 4, pp. 880-895, 2008.

[DES 89] DESTOOP T., "Compresseurs volumétriques", *Technique de l'Ingénieur*, vol. BL2, ref. B4 220, 1989.

[DUB 06] DUBAS F., Conception d'un moteur rapide à aimants permanents pour l'entraînement de compresseurs de piles à combustible, PhD thesis, University of Franche-Comté, 2006.

[ESC 05] ESCRIBANO S. *et al.*, "Study of MEA degradation in operating PEM fuel cells", *3rd European PEFC Forum*, Lucerne, Switzerland, 2005.

[FOU 06] FOUQUET N. *et al.*, "Model based PEM fuel cell state-of-health monitoring via AC impedance measurements", *Journal of Power Sources*, vol. 159, no. 2, pp. 905-913, 2006.

[GLI 05] GLISES R., HISSEL D., HAREL F., PÉRA M.C., "New design of a PEM fuel cell air automatic climate control unit", *Journal of Power Sources*, vol. 150, pp. 78-85, 2005.

[HER 06] HERNANDEZ A., Diagnostic d'une pile à combustible de type PEFC, PhD thesis, University of Technology of Belfort-Montbéliard, 2006.

[HIS 07] HISSEL D., CANDUSSO D., HAREL F., "Fuzzy clustering durability diagnosis of polymer electrolyte fuel cells dedicated to transportation applications", *IEEE Transactions on Vehicular Technology*, vol. 56, no. 5, Part. 1, pp. 2414-2420, 2007.

[HUI 08] HUIZING R., FOWLER M., MÉRIDA W., DEAN J., "Design methodology for membrane-based plate-and-frame fuel cell humidifiers", *Journal of Power Sources*, vol. 180, no. 1, pp. 265-275, 2008.

[JUN 07] JUNG S.H., KIM S.L., KIM M.S., "Experimental study of gas humidification with injectors for automotive PEM fuel cell systems", *Journal of Power Sources*, vol. 170, no. 2, pp. 324-333, 2007.

[LAC 03] LACONTI A.B. *et al.*, *Handbook of Fuel Cells*, John Wiley & Sons Ltd, Chichester, 2003.

[MOÇ 07] MOÇOTÉGUY P. *et al.*, "Monodimensional modeling and experimental study of the dynamic behavior of proton exchange membrane fuel cell stack operating in dead-end mode ", *Journal of Power Sources*, vol. 167, no. 2, pp. 349-357, 2007.

[MOR 03] MORATIN S., Conception, réalisation et modélisation d'un système permettant de contrôler la température et l'hygrométrie d'une pile à combustible de type PEMFC, Mémoire CNAM, Belfort, 2003.

[NAS 00] NASO V., LUCENTINI M., ARESTI M., "Evaluation of the overall efficiency of a low pressure proton exchange membrane fuel cell power unit", American Institute of Aeronautics and Astronautics Inc. (AIAA), 2000.

[ROU 08] ROUSS V., CHARON W., "Multi-input and multi-output neural model of the mechanical nonlinear behaviour of a PEM fuel cell system", *Journal of Power Sources*, vol. 175, no. 1, pp. 1-17, 2008.

[SCH 05] SCHINDELE L., SCHOLTA J., SPÄTH H., "PEM-FC Control using power-electronic quantities", *Electric Vehicle Symposium*, Monaco, 2005.

[STU 08] STUMPER J., STONE C., "Recent advances in fuel cell technology at Ballard", *Journal of Power Sources*, vol. 176, no. 2, pp. 468-476, 2008.

[TEK 04] TEKIN M., Contribution à l'optimisation d'un générateur pile à combustible embarqué, PhD thesis, University of Franche-Comté, 2004.

[TIA 08] TIAN G. *et al.*, "Diagnosis methods dedicated to the localisation of failed cells within PEMFC stacks", *Journal of Power Sources*, vol. 182m no. 2, pp. 449-461, 2008.

[WAH 06] WAHDAME B., Analyse et optimisation du fonctionnement de piles à combustible par la méthode des plans d'expériences, PhD thesis, University of Technology of Belfort-Montbéliard, 2006.

[WAH 08] WAHDAME B. *et al.*, "Comparison between two PEM fuel cell durability tests performed at constant current and under solicitations linked to transport mission profile", *International Journal of Hydrogen Energy*, vol. 32, no. 17, pp. 4523-4536, 2007.

[WIA 00] WIARTALLA A. *et al.*, *Compressor Expander Units for Fuel Cell Systems*, Institute for combustion Engines, FEV Motorentechnik GmBH, Aachen, Germany, 2000.

[WIL 03] WILKINSON D.P., ST-PIERRE J., *Durability, Handbook of Fuel Cells – Fundamentals, Technology and Applications*, John Wiley & Sons Ltd, Chichester, 2003.

[WU 08] JINFENG W. *et al.*, "Diagnostic tools in PEM fuel cell research: part I electrochemical techniques. Part II: physical/chemical methods", *International Journal of Hydrogen Energy*, vol. 33, no. 6, pp. 1735-1757, 2008.

[ZHA 03] ZHAO Y., "Research on oil-free air scroll compressor with high speed in 30 kW fuel cell", *Applied Thermal Engineering*, vol. 23, pp. 593-603, 2003.

Chapter 8

Electrochemical Storage: Cells and Batteries

8.1. Generalities of accumulators: principle of operation

Electrical cells, accumulators and fuel cells belong to the family of electrochemical generators.

Cells are not electrochemical accumulators, as they are not rechargeable. Cells, also known as primary generators, supply an amount of electricity that is predetermined during their manufacture (no charging or preparation is required before use). In effect, they discharge their electrical energy without having the possibility to return to their initial state.

They differ on this point from accumulators, or secondary generators, which can, after discharging, undergo inverse reactions if they are supplied with external electrical energy. Electrochemical accumulators are "reversible" generators, i.e. they can store electrical energy in chemical form and then release it at any moment on demand due to the reversibility of the transformation. They all possess the property of supplying energy using two electrochemical reactions realized at two electrodes that are soaked in an electrolyte. On one of the electrodes, known as the cathode, an oxidizing agent is reduced when electrons arrive at that electrode, whereas simultaneously on the other electrode, the anode, a reducing agent is oxidized and gives up electrons.

Chapter written by Florence FUSALBA and Sébastien MARTINET.

A *battery of accumulators*, is a grouping together of several identical accumulators, using series or parallel mounting.

The *capacity* is the amount of electricity, usually evaluated in ampere-hours (Ah), which a charged accumulator can release during the discharge period.

The capacity of an element is a function of the discharge regime. The nominal capacity of a battery is generally given for a discharge regime of 10 hours (C/10):

– for a higher discharge regime ($I>C/10$) the capacity is reduced.

– for a lower discharge regime ($I<C/10$) the capacity is increased.

The discharge current, in amperes, is evaluated in fractions of the capacity expressed in ampere-hours (for example, C/100).

Example: an accumulator of 100 Ah at C/10 can supply a current of 10 A over 10 hours. Its capacity will be reduced to 80 Ah for a discharge regime with $I = C/5$ whereas the capacity could be increased to 140 Ah for a discharge regime with $I = C/100 = 1$ A.

The capacity of an element is a function of its temperature: it varies according to variations in temperature.

The *faradic efficiency* is the ratio between the amount of electricity used during discharge, Q_D, and the amount of electricity supplied during charging, Q_C. $\eta_q = Q_D/Q_C$.

The *energy efficiency* is the ratio between the amount of energy discharged and the amount charged in Watt-hours. These efficiencies strongly depend on the technologies being considered and the conditions of charging and discharging applied. The energy efficiency is lower than the faradic efficiency because ampere-hours are not stored and released in the same way.

The *self-discharge level* of an accumulator represents the average relative loss of capacity per month in storage for a given temperature. Self-discharge is an internal characteristic of the technology used and is generally given for a temperature of 20°C.

The *internal resistance* of an accumulator is always very low (of the order of a few hundredths of an ohm), almost inversely proportional to the capacity of the element, and in all cases lower for a power device compared to an energy system. This low internal resistance presents a disadvantage: when the two terminals are accidentally linked by a conductor that is itself not very resistant, the total resistance of the circuit is very low; the intensity of the current drawn is considerable. The accumulator, which is short-circuited, is rapidly out of service. The short-circuit test is one of the normalized tests that must be passed for a device to qualify.

In a *floating charge*, the state of charge of the accumulator between the two voltage terminals is limited.

The *lifespan* of accumulators is directly linked to their conditions of operation. For use in buffer storage, the lifespan essentially depends on the number and the amplitude of charge/discharge cycles.

Figure 8.1 presents the general principle of operation of a rechargeable accumulator.

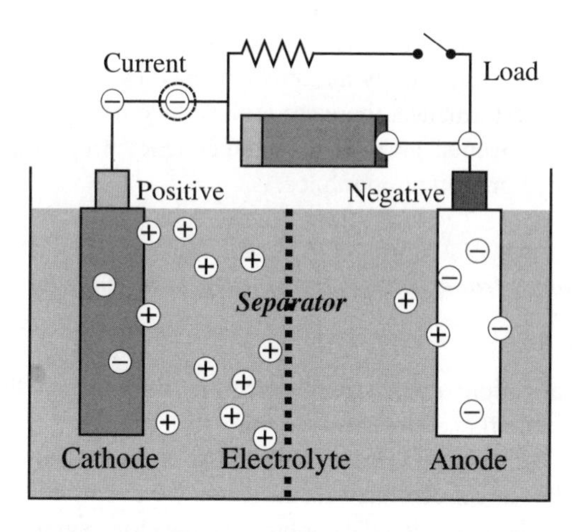

Figure 8.1. *Diagram of the principle of operation of a secondary electrochemical cell (rechargeable)*

So, accumulators and cells are electrochemical systems which release chemical energy generated by electrochemical reactions in electric form, expressed in Watt-hours. The term battery is then used to characterize a set of elementary cells (generally rechargeable). Whatever the technology involved, an accumulator is essentially defined by two values:

– Specific energy, in Watt-hours per kilogram, which is the amount of energy stored per unit mass of accumulator. Also, its specific power, in Watts per kilogram, which represents the power (electrical energy supplied per unit time) that can be supplied by a unit mass of the accumulator.

– Cyclability, expressed in number of cycles, which characterizes the lifespan of the accumulator, i.e. the number of times it can release a level of energy that is greater than 80% of its nominal energy. This value is the value that is most often asked about for portable applications. Therefore, the criteria can change from one application to the next.

8.2. Applications

Electrochemical accumulators are electrical energy storage systems that are used as essential complements for traction applications (cars, scooters, etc.), as well as for renewable energy producing systems (solar, wind power, etc.).

These two types of applications, mobile and stationary, have different battery selection criteria. In addition to criteria associated with the cost, the lifespan (in cycles and on the shelf), the energy and power densities (per unit mass or volume), and the temperature performance, there are criteria linked to non-toxicity of systems, recyclability of components, as well as independence from foreign competition, which are currently of growing importance.

8.2.1. *A global energy framework, using storage to manage electric and transport networks*

With the liberalization of electricity prices in Europe and the creation of the Powernext energy market, electrical energy prices now fluctuate frequently, particularly during consumption peaks. One of the ways to deal with this issue is to store energy, whether in concentrated form (dams, for example) or in relocated form (batteries linked to photovoltaic (PV) solar systems, for example). In addition, the exponential increase in relocated energy production sites, which are mostly intermittent (wind power, solar energy), also has an effect on the variability of prices and leads to a need for storage or complementary production in order to ensure the supply-demand equilibrium. Electricity storage is a good alternative which allows good management of electric networks both in terms of supply and demand and in terms of quality.

In the transport domain, whether it is the current success of hybrid cars, or the development of new generation hybrid plug in devices, or efforts by car manufacturers to make electric vehicles, or else considerations on merging the energy needs of buildings and transport (by Honda, for example) – all these things suggest that electrical energy has an important future in the domain of non-carbon dioxide (CO_2) emitting transport. Therefore, it is becoming a matter of urgency to find technological solutions with high efficiency (at both high power and with high energy capacity) to overcome the current technological obstacles for electrical energy storage for vehicles.

Improvements in battery elements in terms of energy density, security, cost, and the ability to recharge, are major scientific issues which meet the complex specifications for stationary and transport applications.

Regarding hybrid electric vehicles (HEVs), there are different degrees of hybridization, i.e. degrees of electric generator usage during a drive cycle. Hybridization starts with the micro-hybrid version (such as the Citroën C3 *Stop & Go*, which uses lead batteries), which has a low degree of hybridization; continues with the hybrid version of the Toyota Prius (which uses power batteries, Ni-MH) which saves 25% on fuel as the battery is charged by the thermal engine; ending with the latest version with greatest hybridization, known as a Plug-in system, where the battery recharges itself directly from the terminals. Batteries that can combine high energy (range) and high power (acceleration, start-up) are therefore more and more in demand. Professionals in the field agree that it is necessary to stabilize lithium-ion technology to satisfy these specifications.

8.2.2. *Storage technologies in evolution*

Since the invention of the lead accumulator by Gaston Planté in 1859, batteries have continued to improve. Until the end of the 1980s, the two main technologies available on the market were lead accumulators (for starting-up vehicles, supplying secure telephone networks, etc.) and nickel-cadmium (Ni-Cd) accumulators (for mobile tools, toys, security lighting, etc.). Lithium accumulators entered the market early in the 1990s, thanks to Japanese companies. At the time, these accumulators offered a specific energy density (100 Wh/kg), which was greater by a factor of two than that offered by Ni-Cd technology and greater by a factor of three than that offered by lead-acid accumulators. Since then, the performance of lithium accumulators has greatly improved, and in 2008 they attained 200 Wh/kg.

Lithium-ion batteries today occupy more than 70% of the world market, in the field of mobile accumulators, and, from their characteristics and their potential for improvement, they seem to be the most promising technology to overcome limitations in HEV applications and in PV applications. The superiority of the performance of lithium-ion technology stands out in the conclusions of the ASTOR project, which was initiated by European car manufacturers (EUCAR) between 2001 and 2004, and which tested 25 commercial battery systems and technology prototypes. The evaluation of the environmental impact of different battery technologies in the framework of the SUBAT European project between 2004 and 2005 also highlighted lithium-ion technology as the most interesting technology for the HEV application [BOS 06]. The European network INVESTIRE, which is dedicated to the evaluation of storage technologies for renewable energies, has similarly highlighted the suitability of lithium-ion technology for PV applications [THE 02].

8.2.3. *Lithium-ion technology at the heart of HEVs*

The HEV market is expanding, being supported by the politics of pollution reduction and also by elevated petrol prices: 84,000 HEVs were sold in the world in 2004, 205,000 were sold in 2005, and market studies predict the sale of 8 million hybrid vehicles in 2012 in China alone (according to the conclusions of the SUBAT project [BOS 06]). At the moment, commercialized HEVs use an Ni-MH battery with the exception of the Toyota Viitz, which is only on sale in Japan (this is a *Stop and Go* model equipped with a 12V lithium-ion battery). For example, the Toyota Prius and Lexus both use a high-voltage Ni-MH battery. Eighty-three percent of HEVs are sold by Toyota, followed much later by Honda and then General Motors. The Toyota Prius II is a true success from a technical and commercial point of view, marred only by a break in its availability for sale. Nevertheless, we will underline the very low range of the battery (Ni-MH) in the ZEV (*Zero Emission Vehicle*) mode, i.e. the purely electric mode: only 2 km! This suggests that there is still much progress to be made in this domain. This success has led to an increase in interest from all manufacturers in the electric vehicle, due to environmental restrictions, the cost of fossil fuels, and the progress in battery technology.

PSA sells a Citroën C3 *Stop and Start* with a heat engine that cuts out when the car is stopped, in order to reduce consumption and to reduce polluting emissions. This system uses a lead battery, however, it does not allow energy to be recovered when braking, nor does it include boosting of the heat engine during acceleration peaks, due to the limitations of the energy source. The introduction of lithium-ion technology in the hybrid application is expected in the very near future, with a 40-50% share predicted for lithium-ion batteries in 2015. According to the estimates of the SUBAT project, the cost of lithium-ion batteries will fall by 2012, as a result of the growth of the Chinese HEV market (as both producer and consumer) and as a result of the reduction in the cost of activated doped material due to technological innovations.

Plug-in HEVs only exist in prototype form at the moment. We will nevertheless mention in passing two French agents in the electric vehicle domain – Batscap/Bathium (lithium-metal polymer) and Dassault/SVE (lithium-ion polymer) – that are predicted to promote electric vehicles and fleets starting from 2009.

Regarding batteries for hybrid applications, the acceptable depth of charge for a battery without reduction in its lifespan is one of the key parameters. Until now, several battery technologies have competed for HEV applications, namely:

– the graphite/NCA pair, initially developed by SAFT, which is sold by JCS for the hybrid Mercedes S class, and the active hybrid BMW 7 series. JCS will also supply Ford for its first rechargeable hybrid car;

– the graphite/LFP pair (A123, Chinese manufacturers, French Atomic Energy Authority (CEA));

– the $LiMn_2O_4$/graphite pair (notably, Nissan-NEC).

NOTE: NCA = $LiNi_{0.8}Co_{0.15}Al_{0.05}O_2$, LFP = $LiFePO_4$.

In this market, car manufacturers such as Nissan (Nissan-NEC) and Toyota develop their own batteries. In the HEV field, the main players are Toyota, Panasonic, Nissan-NEC, BYD, and LEJ (Mitsubishi), and the main suppliers for accumulator elements are A123, SDI, JCS, Continental, Bosch, and Delphi. The main suppliers of accumulators as complete modules are Sanyo, LGC, JCS, Enerdel, Toshiba, and Murata.

8.2.4. *Lithium-ion technology at the heart of PV solar applications*

For several years, installation of PV panels has been promoted and accelerated by national programs offering financial incentives such as government-subsidized feed-in tariffs for supplying energy to the public grid, particularly in Germany, Japan, Spain, the USA, Australia, France, and in other countries (but each with its particular conditions).

Regarding the PV solar application, the need for energy storage is evident in the case of isolated systems, in order to compensate for the intermittence of the energy source. This being the case, a new function for storage is currently being seriously studied for PV applications connected to the network, as storage would enable smoothing of network peaks in production and consumption, by storing and releasing current during the most pertinent periods (support when production is low, putting off demand depending on the purchase tariffs for electricity from the network and/or feed-in tariffs for the PV electricity).

Lead-acid batteries continue to be principally used in autonomous PV applications, mainly due to their cost and their availability; however, this technology has limits that are slowing the development of these systems. In particular, the lifespans are random, leading to high costs of upkeep (which can reach close to 50% of the costs of the system [RAP 02]).

Traditional storage systems should be reviewed and improved in order to meet the demands for introduction of storage in PV systems connected to the network, which are likely to develop over the next 20 years. As in the case of the HEV, lithium-ion technology seems to be the most promising technology to overcome current limitations in PV applications, and this has been demonstrated by the European network INVESTIRE [THE 02], which is dedicated to the evaluation of storage technologies for systems using renewable energy. The objective in this field is to enter the solar PV market, proposing a new function for storage that has a key property compared to global competition (storage can provide auxiliary services such as: power quality or frequency regulation; it can enter PV market fully integrated with PV modules; it can either store energy for personal use or sale/inject electricity into power grids when requested).

In 2010, the Japanese company, Sharp, will begin running a factory producing lithium-ion batteries that are meant for individual houses that use solar panels. This project will allow electrical autonomy for these houses. The sale price of these batteries will play a major role in the economic viability of this storage solution.

8.2.5. *The French position on the accumulator market*

Examining the production of lithium-ion or Ni-MH accumulators, we note that around 95% are made in Asia and 5% in North America, leaving a very small amount in Europe. Even if France devotes significant effort to the area of hybrid vehicles or renewable energy, while there might be technological innovations at the national level, there would still be a need to use batteries (in vehicles or in PV systems) that are produced outside of Europe by manufacturers who use research and development from private laboratories and universities in their countries. The growing cost of transport, external sources of primary materials (cobalt for lithium-ion, rare earths for Ni-MH), and the current lack of availability of recycling for these devices all demand development of alternative systems at the national level.

At the French level, two battery manufacturers exist: SAFT and BATSCAP. SAFT is essentially positioned on industrial or niche markets with rechargeable Ni-Cd, nickel-metal hydride (Ni-MH), and lithium-ion technologies. BATSCAP is developing lithium-metal polymer technology and also commercializing supercapacitors.

8.2.5.1. *Mobile and niche markets*

The mobile electronics market includes mobile computers, telephones, personal digital assistants (PDAs), pocket games, alarm systems, camping apparatus, military apparatus, and even individual health equipment. In 2005, according to figures from the battery association of Japan (BAJ), the battery market (including primary and secondary batteries (accumulators)) was worth nearly US$5 billion. More than half the turnover (some US$3 billion) originated from the sale of accumulators, used in the design of batteries for mobile computers. In 2006, the rechargeable battery market represented US$7 billion (source: Avicenne 2007).

In volume, the battery market is distributed between cells (73%) and accumulators (27%). However, revenues from accumulators account for 79% of the global revenues from battery sales, of which 41% comprises lithium-ion accumulators.

Today, the rechargeable battery market is distributed between three generations of accumulators: those with a Ni-Cd base, those with a Ni-MH base, and those with a lithium-ion base, whose production is around 100 million units per month.

Lithium-ion technology (which contributes to nearly 80% of the price of a battery) has naturally become the technology for accumulators in mobile computers, ever since the middle of the 1990s. Indeed, compared to other technologies such as Ni-MH, lithium-ion allows more energy to be stored in a smaller volume, while having a longer lifespan.

Mobile electronic equipment, such as phones, computers, and camcorders constitute the principal markets for NiMH and lithium-ion accumulators. After the boom of recent years, these markets must find a growth pattern that is closer to the traditional economy. The shift of applications, from using NiMH to using Li-ion, which is almost complete in Japan, will also happen in Europe. Lithium-ion with a polymer gel electrolyte and flexible casing, introduced in 1999, must continue to make progress in market sectors that demand ultra-flat accumulators, without taking up the dominant position.

8.2.5.2. *Towards new markets*

The increasing popularity of the mobile phone and, on a global scale, of mobile electronics, is accompanied by an increased need for miniaturized energy sources (batteries, cells, etc.). In the short and medium term, the emergence of new miniaturized and interactive products in civil and military domains (autonomous sensors for the development of "intelligent clothing", autonomous medical systems, global positioning system (GPS), onboard sensors, etc.) will accentuate this phenomenon, by opening up new markets.

Therefore, great opportunities for development are offered for production, storage, or energy-recuperation systems. They must reply to the technical challenges presented by next-generation nomad objects, which need new functionalities: freeing the user from the constraints of charging, offering a greater period of use, guaranteeing inviolability of information, collecting available energy from the local environment, and creating an isolated and autonomous energy source, enabling implantation within the human body of biocompatible devices for long durations, etc.

This innovative journey has involved several research and industrial bodies from all over the world. Numerous companies (Samsung, Toshiba, Sony, Motorola, Siemens, etc.) are at the head of the journey, and a large number of start-ups have been created in the USA.

8.2.5.3. *The principal manufacturers*

At the moment, the market for battery manufacturers is essentially – almost exclusively – composed of Asian companies, with the Japanese Sanyo, Matsushita, Sony, and Hitachi, and also the Korean LG, Samsung, and Kokam, as well as the Chinese BYD, ATL, and Lishen.

The Japanese companies rapidly became leaders in the field. Already being manufacturers of mobile equipment, they considered the energy source to be a strategic component in the equipment. This is why Sony, which was not originally an accumulator manufacturer, decided in the 1980s to dedicate considerable resources to developing the technology and making it industrializable. In February 1992, Sony announced, to general surprise, the immediate launch of industrial manufacture of lithium-ion accumulators. These first accumulators offered limited performance (90 Wh/kg). Since then, they have improved notably (from 160 to more than 200 Wh/kg in 2008), thanks partly to technological progress (reduction of unnecessary weight and volume in accumulators), and partly to the optimization of the performance of materials.

In the case of the mobile electronics market, it is the Japanese, Koreans, and Chinese that dominate. However, in this field, the market remains open to new applications requiring innovative accumulators (rapid recharge, printed battery, etc.), often at a high price (medical applications).

The lithium-ion battery sector has been in turmoil since 2004, with the arrival of new manufacturers that have emerged from the USA (A123Systems, EnerDel), Korea (Kokam, LG Chem), or China (MGL, B&K, HYB, BYD) as well as the big world players in this field (Panasonic, Nec, Hitachi, Saft). Japanese car manufacturers (Toyota and Nissan) invest in battery companies (such as Panasonic, but also new manufacturers such as GS Yuasa, Lithium Energy Japan) in order to master this strategic component for the hybridization of vehicles.

For lithium-ion batteries based on $LiFePO_4$, their elements are sold by A123 (A123 material), Saft (HydroQuebec material, Phostech), Valence Technology (Valence material) or Chinese manufacturers (Chinese materials). These cells have energy densities between 90 and 110 Wh/kg, with products currently being created for power applications rather than energy applications.

At the national level, Bolloré has opened its Batscap factory (lithium-metal-polymer, LMP, technology) near Quimper, dedicated to its electric car (Bluecar); Johnson Controls-Saft has just opened its factory in Nersac, for manufacture of lithium-ion power elements in France, also destined for use in vehicles. Further afield, Saft is offering NiCd, NiMH and lithium-ion batteries (being the first manufacturer in the world of lithium primary batteries for electronics and defense) for all professional and mobile industrial applications: Saft's batteries are used in high-performance applications, notably within industrial infrastructures and processes (stationary applications and security lighting), in air and rail transport (premier manufacturer in the world of Ni-Cd batteries), in defense and in space (second manufacturer in the world, and first in Europe). Saft is the premier manufacturer in the world of Ni-Cd batteries for industrial use and of lithium primary batteries for a variety of applications. The Saft group is also the premier European manufacturer of batteries with specialized technologies for defense and for space.

At the European level, Johnson Control, with the old Varta car batteries, supplies all the European car manufacturers. It has six manufacturing sites in Europe and mainly delivers to manufacturers such as BMW, DaimlerChrysler, Ford, PSA, and Volkswagen, as well as to Bosch and to major resellers such as Carrefour. Its joint venture with Saft allows Johnson Control to manufacture batteries (Li-ion) for the new-generation HEVs.

At the world level, Johnson Controls (USA) is the main global supplier of lead batteries. Johnson Controls is positioned as a world leader with experience of automobiles, building regulations, and electrical supply solutions.

8.2.5.4. *The academic environment*

For almost 15 years now, the CEA has been developing electrochemical storage of energy. Within its laboratory for innovation with new energy technologies and nano-materials (LITEN), which is mainly based in Grenoble, its research and development activities are mainly concerned with solar technologies, hydrogen, and nano-materials, and are aimed at three markets: buildings, transport, and nomad electronics. LITEN uses a line of prototypes that enable it to design and realize lithium accumulators for HEVs, for coupling with solar power and with the electric network, for new applications requiring innovative accumulators (rapid recharge, etc.), often at high level (medical applications, thin or printed batteries, etc.). This line is based on several material platforms (synthesis equipment, mechano-synthesis platform, etc.).

Among the French laboratories working in this field, LRCS (Amiens) is especially strong, especially in the area of materials. We can also cite IMN (Nantes), LGMPA (Nantes), LEPMI (Grenoble), ICMCB (Bordeaux), but also CEMES (Toulouse), etc.

The European network of excellence Alistore (*Advanced Lithium energy STORage systems*) which was started in 2004 and coordinated by a French laboratory (LRCS Amiens) groups together 14 laboratories and 12 industrial companies to work on the development of new storage systems that are mainly based on lithium.

8.2.5.5. *American and Japanese orientation regarding energy storage*

The specifications set out by the DoE (US Department of Energy) regarding batteries for HEVs have the following objectives:

– energies of around 1-2 kWh, with a power/energy ratio of greater than 15;

– for the HEV *Plug in* application the objective is around 5-15 kWh, with a power/energy ratio of between 3 and 10;

– and for an electric vehicle the objective is greater than 40 kWh, with a power/energy ratio equal to 2.

The DoE's total HEV budget was US$94 million (€66 million) in 2008, of which US$48 million (€34 million) was for batteries. The DoE has invested in fundamental applied research via the BATT program, which involves universities and national laboratories, and in the development of the battery pack via the USABC program together with car manufacturers.

The technologies developed either use positive cathode materials, such as lamellar oxides or high voltage, or $LiFePO_4$ coupled with graphite as negative material, or they use titanium oxides to replace the negative electrode. Other electrochemical systems are being developed, for instance the lithium-sulfur system.

In 2008, the cost of technologies was €28/kW, which is twice as much as the target for a commercializable version. For the *Plug in*, the target energy costs were €350/kWh (PHEV10, with 10 mile range), and then €210/kWh (PHEV40, 40 miles). The current state is in the order of €700/kWh for this application. The corresponding demands in specific energy terms are 100 Wh/kg in 2012 and 150 Wh/kg in 2015. Among the technologies envisaged to attain these values, 5 V cathodes and lithium alloy anodes (Si-C for example) or titanate anodes are cited.

NEDO (in Japan) is positioning itself in the same domain, but with more ambitious objectives. Regarding storage for applications connected to the network, it predicts the need for more than 90 GW of battery capacity in order to smooth the current and improve current quality, with an objective of €100/kWh for a 20 year lifespan. At the moment, Enax, Mitsubishi, and Hitachi are working on this.

Regarding vehicles, specifications in 2015 will require 100 Wh/kg, 2 kW/kg, a 1-year lifespan, and €270/kWh at the module level, or 200 Wh/kg and 2,500 W/kg at the cell level. The objectives in 2030 (module) are 500 Wh/kg and €33/kWh! NEDO mentions that a technological breakthrough is required in order to attain 700 Wh/kg and 1,000 W/kg in 2030.

8.3. Technological histories: lead, Ni-Cd, Ni-MH...then lithium ion

8.3.1. *Lead/acid*

8.3.1.1. *Principles and technology*

Lead battery technology (invented by Planté in 1859) is the most widespread electrical energy storage technology and has evolved considerably since then. Its cost : energy ratio being by far the lowest, which should keep it in service for many years to come. There are two types of lead batteries, defined according to the nature of the electrolyte.

The liquid electrolyte battery offers an excellent lifespan but demands regular and frequent maintenance. Some devices allow this to be reduced to intervals of 200-250 discharge cycles or the equivalent of once a year.

The lead/acid battery is an accumulator that is based on lead and on sulfuric acid, formed from a negative plate made of spongy lead and a positive plate of lead oxide (Figure 8.2). The majority of electric vehicles use this type of battery. The separators, which are necessary to isolate the positive and negative plates to prevent a short circuit, are simple cellulosic sheets in the traditional batteries. In high-performance batteries, the separators are micro-porous which prevent any risk of a short circuit.

The voltage at the terminals of an element in the lead accumulator is around 2 V. Its value varies between 1.7 V and 2.4 V depending on the state of charge in normal operating conditions.

The following electrochemical reactions take place at the electrodes:

– at the anode (oxidation):

$$Pb(s) + HSO_4^-(aq) \leftrightarrow PbSO_4(s) + 2e^- + H^+ \ \varepsilon^0 = -0.356 \text{ V};$$

– at the cathode (reduction):

$$PbO_2(s) + HSO_4^-(aq) + 3H^+ + 2e^- \leftrightarrow PbSO_4(s) + 2H_2O(l) \ \varepsilon^0 = 1.685 \text{ V}.$$

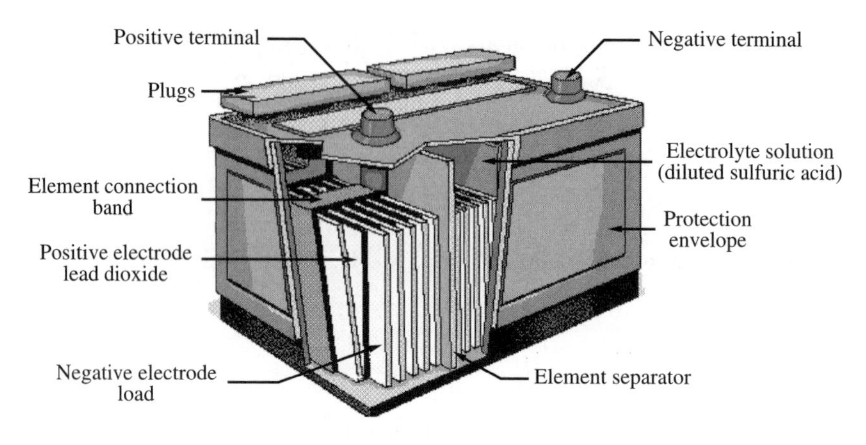

Figure 8.2. *Diagram of a lead/acid battery*

The zero-maintenance battery exists with two types of electrolyte: silica gel or non-woven fiber membrane. Generally equipped with a regulation valve and allowing recombination of gases, this battery does not require maintenance. With this technology, the battery and the charger form an integrated system that ensures optimal usage – note that these batteries generally have a weak tolerance for overcharging. The zero-maintenance lead battery has an energy density that can vary from 35 to 50 Wh/kg. The concept of "bipolar" batteries avoids the conventional assembly of positive and negative grates, thus enabling performances to be superseded. These batteries have a lifespan of around 500 cycles at 80% discharge.

Numerous manufacturers exist across the world and almost all are participating in the battery improvements advocated by the *Advanced Lead-Acid Battery Consortium* (ALABC).

We can distinguish four major "subcategories":

a) *Start-up batteries* (with grating), manufactured in many countries of the world; these models are easy to find locally. They are designed to be kept permanently charged and to deliver strong currents quickly. They self-discharge. They are not very expensive (€0.2/kWh released) and are widely available.

b) *Drive batteries* (with flat plates), these are derived from those that are used in electric vehicles and more traditionally on autonomous carts (railroad, elevators). They are designed to be charged and discharged on a daily basis. Therefore, they can function at a low state of charge. They are moderately expensive (€0.5/kWh released).

c) *Batteries with gel electrolyte* (with flat plates), these are also mainly from the traction area, often with low capacities of the order of 100 Ah. They do not need any maintenance, lasting from 3 to 5 years for a cost in the order of €0.6/kWh (released). They are frequently used in small professional installations (radio communication), especially in moving ones (shining buoys). There also exist "waterproof" batteries where the liquid electrolyte is absorbed and permanently kept in synthetic tissue pockets.

d) *Stationary batteries* (with tubular plates) are the result of security supply technology (telecom relays, etc.). They are designed to be permanently charged with a weak current (known as a floating current since it keeps the battery at a constant voltage), have a weak self-discharge (1 or 2% per month), and supply almost all their energy when the need arises. Nevertheless, they are able to be cycled, and to remain for a few weeks at a relatively low state of charge. Although they need a minimum amount of maintenance (adjustment of the water level) and cost €0.5/kWh, they are robust (8-12 years lifespan) and therefore are systematically used to equip battery stations of more than 300 Ah.

In practice for all the lead technologies it is not possible to discharge below 20% of the battery capacity. Otherwise, sulfation (see the following section) leads to a loss of capacity and an increase in the internal resistance and therefore a lowering in the energy released.

8.3.1.2. *Controlling the state of charge*

During discharge, the concentration of sulfuric acid decreases. In contrast, some sulfuric acid is formed during charging. The density of the electrolyte is measured to verify the state of charge, which enables the acid concentration to be determined. After charging is finished, if the passage of current is prolonged, the hydrogen and oxygen resulting from the decomposition of water will end up being released in gaseous form at the electrodes by electrolysis. If the discharge takes too long, or if the concentration of the sulfuric acid is not monitored, the acid can attack the plates, creating lead sulfate which cannot be destroyed. The accumulator can become useless. Therefore, it is important to monitor the state of charge or discharge of a lead battery in order to maintain a good state, as prolonged operation in either direction will lead to the definite destruction of the accumulator.

8.3.1.2.1. Self-discharge and lifespan

The rate of self-discharge depends on the type of materials used (lead alloys, separators, etc.). The rate of self-discharge is of the order of 10% per month, for antimonious lead plates (this alloy aims to improve the mechanical performance). It is of the order of a few percent per month at 20°C for soft lead with a low amount of antimony or calcium lead, but the elements are more fragile. Self-discharge varies rapidly with temperature and to a first approximation it is possible to say that it doubles in value with every 10°C, following the law of Arrhénius.

By limiting the daily depth of discharge (<15% NC) and the seasonal depth of discharge (<60% NC), the lifespan of lead accumulators can be estimated as 6 or 7 years, provided they are protected against overcharging. NC represents the nominal capacity of the battery.

8.3.2. *Ni-Cd*

Ni-Cd technology, which is historically one of the oldest technologies, dating from 1899, is today well known. The anode is chemically impregnated with nickel and the cadmium cathode is molded in plastic over a steel substrate. The separator is made of non-woven fibers and the electrolyte is an alkaline liquid. The cells are mounted in a polypropylene case.

The Ni-Cd accumulator is made of electrodes of cadmium and NiOOH in an concentrated potash electrolyte. The discharging reactions are as follows:

$$Cd + 2OH^- \rightarrow Cd(OH)_2 \text{ (solid)} + 2\, e^-$$

$$NiOOH \text{ (solid)} + e^- + H_2O \rightarrow Ni(OH)_2 \text{ (solid)} + OH^-$$

In the case of industrial batteries for electric traction, for example, the battery is cooled by a water or air system, which needs regular maintenance every 50 to 100 discharges. The energy density is 55 Wh/kg and the specific power has reached 100 to 135 W/kg.

The nominal voltage is 1.2 V, but it varies between 1.15 V and 1.45 V depending on the state of charge.

Ni-Cd accumulators can take a large number of complete discharges without destruction, and can be used in the cold – they can operate over a great range of temperatures. The elements are, by construction, more robust and less heavy than the equivalent lead elements, they can more easily accept overcharging or deep discharge, and they need reduced maintenance, which gives them a long lifespan.

However, there are also disadvantages: their cost is around five-times greater than for lead technology, they have poor energy efficiency for charging/discharging (faradic efficiency = 70%), the rate of self-discharge is higher than for lead accumulators (>15%) and they exhibit a "memory" effect, which is linked to the negative cadmium electrode (it is preferable to empty the battery completely before recharging).

Compared with Ni-MH, Ni-Cd can manage more significant (10-times more) discharge current peaks, but its natural self-discharge is more rapid than for Ni-MH. Cadmium is very polluting, which threatens to eventually render this technology obsolete. Ni-Cd batteries have today been overtaken in terms of autonomy: they were overtaken around 1990 by Ni-MH batteries, which are themselves competing against lithium-ion batteries today.

8.3.3. *Ni-MH*

Ni-MH battery technology has been shown to be high performing at ambient temperature. It is practically non-polluting compared to technologies such as lead-acid or Ni-Cd. The system works in a similar way to Ni-Cd technology.

The nominal voltage of an elementary accumulator of this type is 1.2 V. Its specific energy is 40% greater than that of Ni-Cd and its memory effect is very low, due to suppression of the negative cadmium electrode.

The end of charging is characterized by a very weakly negative variation in the charge voltage ($\delta v/\delta t$). It is this threshold that is detected by good quality automatic chargers in order to stop the charging.

The original design for a bipolar battery was introduced in the USA. The module consists of a stack of wafer cells. The technology exploits the chemical, thermal, and electric properties of a thin, conducting, plastic film. Current flows across the cell interfaces, perpendicularly to the electrodes. These electrodes are manufactured using a plastic molding technique, to give a higher specific capacity.

A variant of the Ni-MH battery does not contain cobalt in the positive electrode, which is considered by some to be of no use in block batteries. Cobalt would essentially serve to control the release of oxygen in case the elements went above the necessary voltage.

8.3.4. *Nickel-zinc*

Accumulators based on zinc, such as nickel-zinc (Ni-Zn) or silver-zinc (Ag-Zn) devices, operate in an alkaline environment with easily recyclable compounds whose primary materials are easily acquired. They offer a range of applications from consumer products to use in military or space systems.

They have the advantage of being less expensive than Ni-Cd and of having a usage voltage that is 25% higher. Conversely, they have poor resistance to cycling (around 600 to 1,000 cycles). The electrochemical reaction linked to Ni-Zn technology is as follows:

$$2NiOOH + Zn + H_2O + KOH \leftrightarrow 2Ni(OH)_2 + K_2Zn(OH)_4$$

The Ni-Zn accumulator is one of the rare systems using an aqueous electrolyte that can operate at voltages that are higher than the decomposition voltage for water. This is linked to the thermodynamics of zinc, which also has a high theoretic capacity (820 Ah/kg). Also, the abundance of zinc, its low cost, and its lack of toxicity are all assets. This explains the numerous works that have been carried out with this system, which have all hit problems with the cycling performance of the zinc electrode in alkaline conditions: the formation of dendrites and massive redistribution of zinc.

In order to counter these difficulties, multiple solutions have been proposed. We will especially consider those relating to accumulators with an immobilized electrolyte, for which the principal objective is to limit the solubility of the zincates, which are products of the reaction of zinc in alkaline conditions. In order to do this, solutions with a weak concentration of potash (less than 5 M) are used, with additives such as fluoride, phosphate, carbonate, etc., which form insoluble compounds with zinc. We also add lime to the active mass of the electrode to form insoluble calcium zincate, which is stable when the concentration of potash is less than 7 M.

In order to delay the increase in dendrites, zinc exfoliations, which form during charging, it is usual to use multiple layers of micro-porous separators, or even ion-exchanging membranes. Although these different solutions improve the cycling duration of the accumulators (around 200 cycles), they penalize the Ni-Zn system by increasing the internal impedance, reducing the specific powers and energies, and putting a strain on the cost, notably by having multiple layers of micro-porous separators or ion-exchanging membranes.

During the 2000s, the American company, Evercel, developed Ni-Zn accumulator production on an industrial scale, initially in the USA, but this activity was transferred to China. The technology is based on the use of low concentration alkaline solutions, with calcium zincate as the active mass (US patent no. 5863676) and multiple layers of separators, which have never allowed operation beyond 300 cycles. Other American companies are developing Ni-Zn technologies: Evionyx, via its Xellerion subsidiary, is proposing accumulators that include an ion-exchanging membrane (US patent no. 7119126). PowerGenix is developing Ni-Zn accumulators for mobile applications in the Sub c format, with a modified alkaline electrolyte containing fluorides, borates, phosphates, etc., which reduce the solubility of the zincates (US patent no. 2006207084).

Therefore, from now on, Ni-Zn is both an energy and power system, with performance superior to that of Ni-Cd and Ni-MH. It can accept elevated charging and discharging regimes. Its nominal voltage is 1.65 V and its cycling lifespan is equivalent to that of Ni-Cd; however, its self-discharge and effective memory are less. Ni-Zn is a robust accumulator, which is reliable and can function in a zero-maintenance mode (watertight). Its fabrication is more economical than for other alkaline accumulators (Ni-Cd and Ni-MH). It does not contain any heavy metals, and is easily recyclable at the end of its lifetime. However, this technology must still be tested as no product has yet become widely available.

8.3.5. *Sodium-sulfide (Na-S)*

The Na-S cell is presented in the form of a cylinder, which is filled with sodium, while sulfur makes up the perimeter (see Figure 8.3). The two elements are separated by a ceramic electrolyte (beta alumina). The entire system is mounted in a steel composite case. The cell operates at 300°C. The specific energy is of the order of 100 Wh/kg for a specific power of the order of 230 W/kg. These characteristics represent triple the performance of Ni-Cd.

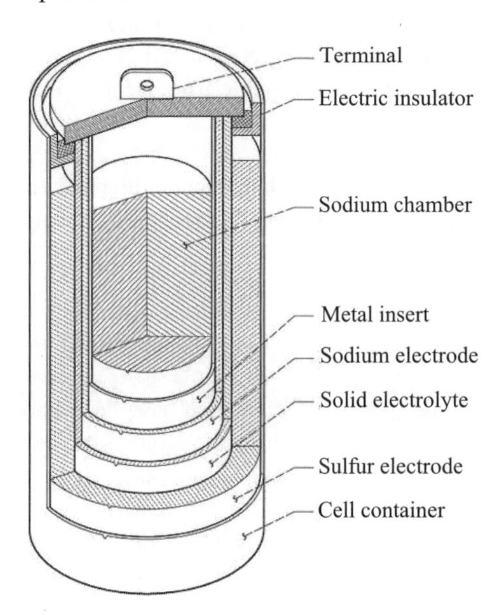

Figure 8.3. *Diagram of a sodium sulfide battery*

Taking into account the potential high risks, this technology is limited to certain stationary applications (storage of intermittent energies, etc.; see Chapter 3).

8.3.6. *Redox*

Redox batteries are batteries with electrolyte circulation in which the chemical compounds are in solution. Several associations with bromine can be envisaged: with zinc, sodium, and more recently, with sodium poly-sulfide. The electrochemical reaction across a membrane in the cell is reversible (charge and discharge). By using large reservoirs and coupling numerous cells, great quantities of energy can be stored and released. For example, Regenesys Technologies in England has made a storage system using this process in 2003, which has a range of 15-120 MW, but this technology is now dominated by electrochemistry that is entirely based on vanadium. The global efficiency of the storage is around 75%.

8.3.7. *The Zebra system*

The sodium-nickel chloride battery ($Na-NiCl_2$), operates at high temperature (300°C) at a nominal voltage of 2.58 V per cell. The negative electrode is made of liquid sodium whereas the anode is a nickel chloride. The sodium is separated from the chloride by a ceramic electrolyte which allows sodium ions to pass during charging and discharging. The system thing is mounted in a metallic case which acts as a negative terminal. The battery can generate from 8.5 to 30 kWh with a specific energy of 85 Wh/kg for a power which varies from 72 to 130 W/kg depending on the cooling mode. Its lifespan is 1,000 cycles at 80% discharge. Still in the research and development stage, its operation at high temperature and the risks in case of breakage limit its large-scale use. However, this technology would be more secure than Na-S.

8.3.8. *Zinc-air*

Metal-air batteries (Figure 8.4) are products that have been known for decades in various forms, from the small "button" cell to the large cubic cell which is used to supply an electrified livestock barrier for several months.

Products derived from these use a suspension of a metal in an electrolyte, such as, for example, powdered zinc in a potash solution, which circulates in a beam of tubes that is externally equipped with air electrodes. This system can reach current densities that are compatible with the power required to drive a vehicle. However, the metal-air (zinc or aluminum) technology requires many precautions to be taken during manipulation. However, its rapid recharging capacity makes it attractive for applications such as the electric float vehicle.

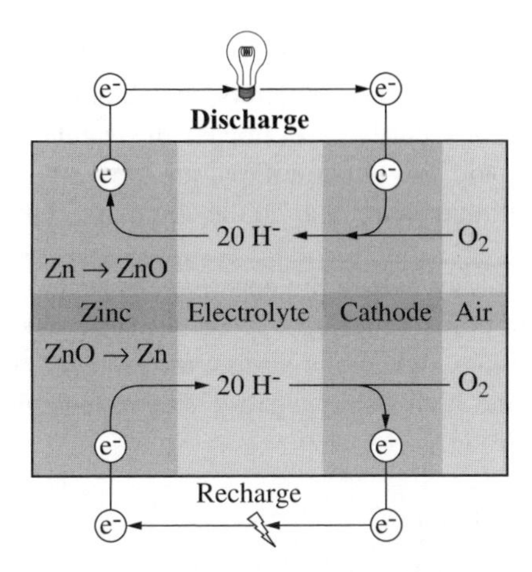

Figure 8.4. *Diagram of a metal-air battery (source ESA Technology website)*

Therefore, a future possibility is an electric vehicle equipped with a Ni-MH hybrid battery, which will ensure current peaks and recuperation of energy during braking, linked to an air-zinc cell, which brings the electrical energy required for a long range (1,000 km or more). While in 2008, Toyota claimed it was working seriously on this type of system for future electric vehicles with long range.

The success of such an option depends on numerous parameters, such as the choice of metal or alloy in circulation, the control of the air electrode (which falls back on studies of the fuel cell), the miniaturization of all hydraulics and elements of the cell, the thermal and electrical control of the system, the possibility of electrically recharging the system in order to avoid excessive changing of used electrolyte that requests filling up with new suspensions. Rechargeable metal-air batteries that are being developed use metals with high energy densities when they are oxidized, such as zinc or aluminum. Their cathodes or air electrodes are often made of a porous carbon structure or of a metallic grill impregnated with the appropriate catalyst. The electrolyte is a good conductor of OH^- ions, whether liquid or solid (polymer membrane saturated in KOH). These rechargeable batteries have a lifespan of around 100 cycles and an efficiency of around 50%. The zinc-air cells can produce up to 450 Wh/kg whereas aluminum-air cells can potentially produce specific energies that are greater than 550 Wh/kg. Using low cost materials with high energy densities will only be truly attractive once their recharging capacity has been mastered.

8.3.9. *Lithium*

Lithium is anticipated to be the mainstay for long-term objectives of the USABC. *Lithium-ion* accumulators have already been replaced, for some applications, by *lithium polymer* accumulators, which deliver a little less energy, but are safer. The problems of security (fire) in case of overcharging, over-discharging, or short-circuiting, remain an important parameter to be controlled before the introduction of a new lithium technology to the commercial public.

Lithium metal technology, where the negative electrode is composed of metallic lithium (a material that poses significant security problems), is to be distinguished from *lithium-ion* technology, where the lithium remains in ionic state, due to the use of an insertion compound both at the negative electrode (usually made of graphite) and at the positive electrode.

8.3.9.1. *Lithium metal generators*

During discharge, the anode (the negative electrode made of metallic lithium) is the seat of oxidation: lithium ions cross the electrolyte and go to the cathode (positive electrode) to undergo reduction by mixing with a specific material (a host material denoted by H, see Figure 8.5). The electrons thus produced supply the exterior circuit with energy. During charging, Li^+ ions make the opposite journey, with electrons being supplied by the exterior circuit.

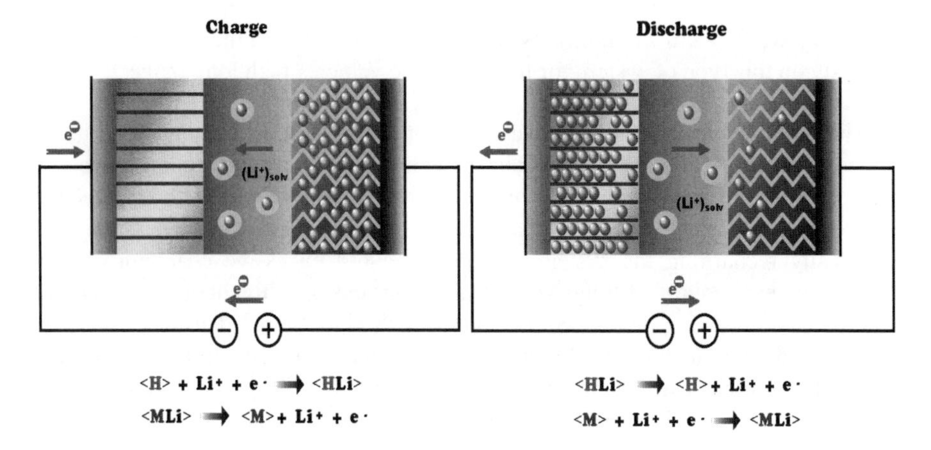

Figure 8.5. *Diagram of a lithium-ion cell or element when charging and discharging*

A *lithium-ion* accumulator is different from the preceding type at the level of the negative electrode: pure lithium is replaced with carbon (graphite) in which lithium can enter. The main asset of this system is the absence of direct contact between lithium and the electrolyte and a consequence of this is an improved chemical inertia for the system. The potential of this electrode is very slightly greater than that of the pure lithium electrode and the mass of the electrode is generally greater, thus reducing its specific capacity from 3,828 Ah/kg for pure lithium to 340 Ah/kg for the LiC_6 compound.

The lithium-ion concept uses two materials that allow the reversible exchange of lithium ions. The anode is a thin layer of graphite, in which atoms of lithium have been inserted (LiC_6). For the positive electrode, a lithium oxide of a transition metal such as $LiCoO_2$ can be used. The liquid electrolyte is usually a hexafluorophosphate of lithium ($LiPF_6$) mixed with a solution, which is a mixture of carbonates. The cell operates at a voltage of around four volts. The most typical potential of a lithium-ion cell is 3.7 V.

The electrolyte in a lithium generator can only be a purely ionic conductor and therefore is a good electronic insulator. This electrolyte is, for the majority of cases, a liquid (a lithium salt: $LiAsF_6$, $LiPF_6$, $LiClO_4$, etc., dissolved in an organic solvent or mixture of solvents such as propylene carbonate, ethylene carbonate, dimethoxy-ethane, etc.). It can also be solid: an organic compound based on polyethylene oxide or an inorganic compound based on amorphous lithium borate, for example. Polymer electrolytes belong to the family of solid electrolytes. These are complexes between a polymer, such as polyoxyethylene (PEO), and a salt of lithium, such as $LiClO_4$. In gel polymers, the lithium salt is dissolved in an organic solvent, such as ethylene carbonate, while being immobilized in a polymer such as polyacetonitrile (PAN) or copolymers PVdF-HFP. These materials have significant ionic conductivities, of the order of 10^{-5} S/cm to 10^{-3} S/cm, even if their performances are generally below those of liquid electrolytes.

8.3.9.2. *Lithium-iron phosphate*

Lithium-iron phosphate is a variant of lithium-ion technology. A positive electrode made of the active material $LiFePO_4$ is associated with a negative graphite electrode in these accumulators, which present a slightly weaker voltage (3.2 V), but offer greater security, a lower cost, and greater cycling stability.

8.3.9.3. *Lithium-metal polymer (Li-po)*

This technology uses a negative lithium metal electrode and a positive electrode based on an oxide of vanadium. The electrolyte is a polymer based on POE (ethylene poly-oxide) and on a lithium salt, which functions optimally between 60°C and 80°C. This technology is realized using extrusion, and is developed by BATSCAP (France) and AVESTOR (Canada). Currently expensive, this new technology is promising for the future and should, eventually, become less expensive to make than traditional lithium-ion technology.

Professor Michel Armand is the designer of the lithium-polymer cell. Technically, the lithium-polymer accumulator is viable and can cost between US$100 and US$150/kWh. It operates very well between 60°C and 150°C without security problems. Research is going on to reduce this temperature a bit, but "this battery will not function at –20°C". However, at the right temperature it requires very little energy. The loss is between 10% and 20% after a week. For example, the lithium metal polymer battery developed by Batscap and designed for an electric vehicle application can reach an energy density of 110 Wh/kg (110 Wh/liter). It functions at 90°C (internal temperature).

Lithium-ion technologies should see interesting developments provided the high-temperature risks can be solved. For example, lithium-carbon works well at low temperatures but, beyond 60°C, it can heat up due to the organic electrolyte used. There is no danger for small batteries, but it cannot exceed a certain critical mass.

Technology	Specific energy density (Wh/kg)	Energy density per unit volume (Wh/L)	Voltage of an element (V)	Charge time (h)	Peak power (W/kg)	Number of cycles (ch./dch.)	Self-discharge per month (%)	Opera-ting temp (°C)
Lead acid	30-50	75-120	2	6-12	700	400-1,200	5	–20 to 60
Ni-Cd	45-80	80-150	1.2	1-2	?	2,000	>20	–40 to 60
Ni-MH	60-120	220-330	1.2	2-4	900	1,500	>30	–20 to 60
Li-ion	100-200	300-550	3.7	2-4	1,500	500-1,000	<10	–20 to 65
Li-ion polymer	100-130		3.7	2-4	250	200-500	<10	0 to 60
$Na-NiCl_2$ (ZEBRA)	120	180	2.6	Unknown	200	800	>100 (12%/j)	Unknown
Ni-Zn	70-80	120-140	1.65	Unknown	1,000	>1,000	>20	Unknown

Table 8.1. *Battery technology specifications*

Techno.	Lead acid	Ni-Cd	Ni-MH	Li-ion	Li-ion polymer	Zebra	Ni-Zn
Cost comparison (2004)	$25 (6 V)	$50 (7.2 V)	$60 (7.2 V)	$100 (7.2 V)	$100 (7.2 V)	Unknown	Unknown
Cost per cycle (2004)	$0.10	$0.04	$0.12	$0.14	$0.29	Unknown	Unknown

Table 8.2. *Battery technology costs*

System	Advantages	Disadvantages
Lead	Low investment	High mass sudden death
Nickel-cadmium	Robustness, reliability	Higher cost than lead Mean specific energy
Metal Nickel - hydroxide	Very good energy density	Cost of base materials
Nickel-zinc	Reasonable intial cost	Very low lifetime
Na-$NiCl_2$	High specific energy	Works at 300°C Lack of power
Lithium-ion	Excellent specific energy	Not mature for industrial application

Table 8.3. *Advantages and disadvantages of the principal electrochemical systems (source: Forum Energies, 2001, organized by the ADEME, Anne de GUIBERT, SAFT, Battery evolution: applications and perspectives)*

8.4. Application needs

8.4.1. *HEV and electric vehicle applications*

Needs differ depending on the type of vehicle:

– HEVs need power (autonomy is not required), the power : energy ratio, P:E, is greater than 10.

– *Plug in* HEV and electric vehicles need power and energy (as autonomy is required), the P:E is between 3 and 5.

As a result, storage solutions differ depending on the type of vehicle:

– HEVs: Ni-MH batteries are the most profitable at the moment, but they are at their maximum performance. Lithium-ion is in the process of being integrated, although it is not currently competing with Ni-MH.

– *Plug in* HEV: lithium-ion technology is solely used in this application. In particular, lithium-ion technologies with reinforced security such as $LiFePO_4$/graphite or $LiMn_2O_4$/graphite are predicted to appear quickly;

– electric vehicles: same remark as for *Plug in* hybrid vehicles, with particular attention to the Li-po technology promoted by Bolloré, via Batscap and its BlueCar vehicle.

8.4.1.1. *Needs*

Regarding transport, two requirements are currently distinguished:

– a traction battery for electric vehicles that has a high energy density with a cost of less than US$150/kWh;

– a power battery for hybrid vehicles with a minimum power of 3 kW/kg and high cyclability.

In both cases, the storage solution must have a lifespan of greater than 10 years, and total security.

Despite the fact that lithium batteries are threatening to dominate the market, lead batteries continue to compete: several companies aim to improve this technology for electric and/or hybrid vehicles by making it more enduring while keeping its low cost. Advanced lead batteries come in three categories:

– bipolar lead batteries (developed notably by Effpower);

– compressed lead batteries resulting from a lead start-up technology (developed by Exide);

– hybrid batteries combining lead batteries and a supercapacitor in the same accumulator (the concept of the Ultrabattery developed by CSIRO – *Commonwealth Scientific and Industrial Research Organization* – in Australia and the Japanese battery, Furukawa).

Out of these three categories, we will underline the integration of the Ultrabattery in the hybrid vehicle, the results of which show that this newly designed battery can compete with Ni-MH. The Ultrabattery concept links the lead battery and a supercapacitor due to replacement of part of the negative electrode by a carbon electrode, which creates an asymmetric supercapacitor along with the positive electrode made of lead dioxide. This allows the battery to accept and deliver strong currents such as those encountered in the HEV application.

Similarly, numerous manufacturers (General Motors, BMW, PSA Peugeot Citroën, and Continental) are working on the hybridization of lead batteries and supercapacitors; firstly, in order to answer the increase in electric needs onboard vehicles, and secondly, as a potential solution for the HEV.

8.4.2. *PV applications*

PV systems consist of solar panels, a battery for energy storage that ensures a quasi permanent energy supply no matter what the level of sunshine, and a regulator that enables management of the energy between the module, the battery, and the charge consumer. The reader may refer to Chapter 3, which details the storage technologies for this type of application.

8.4.3. *Mobile electronic applications*

8.4.3.1. *Needs*

A mobile phone consumes around 1 W in conversation, and 50 mW on standby. A mobile computer needs around 10 W. In addition to demands on power and price, it is necessary to be able to recharge these apparatus in a simple and rapid way and above all to ensure that they have a more significant range than they do at the moment. All the more so as the consumption of these apparatus is likely to increase in the years to come with the increase in functionality, for example, the use of the internet on mobile phones.

The key issue for battery manufacturers (and indirectly for computer manufacturers) is the need to further secure their products to avoid unfortunate disappointments, such as Sony having to recall 4.1 million batteries (2006); followed by Apple recalling of 1.8 batteries in the same year.

8.5. Focusing on lithium-ion technologies

8.5.1. *Principle*

At the moment, electrochemical lithium-ion accumulators are rechargeable systems that can attain higher stored energy densities (per unit mass or volume) than any other existing rechargeable systems. This type of accumulator currently allows storage of 300-550 Wh/liter and 100-200 Wh/kg, for a nominal voltage of 3.7 V, with low self-discharge values (5% to 10% a month) and a wide operating temperature range (–20°C to 65°C). The current technology can reach specific energies of 200 Wh/kg and 55 Wh/liter.

These lithium-ion accumulators use the reversible electrochemical insertion of lithium ions into two materials, with different potential values. These two active electrode materials are a mixed oxide with a cobalt base (generally $LiCoO_2$ or $LiNi_{0.8}Co_{0.15}Al_{0.05}O_2$) at the positive electrode, and graphite at the negative electrode. In 2008, Sony proposed a new lithium-ion polymer technology, produced in Singapore, with a high specific energy (Apelion, 241 Wh/kg, 535 Wh/liter) based on $LiCoO_2$ and graphite, using a gel electrolyte. According to Sony, the technological limit seemed to have been reached.

Current works on active electrode materials (positive or negative) for the lithium battery are particularly interested in researching the compositions and physico-chemical characteristics that could increase the energy density and power density. In addition to technical performance requirements, other criteria that are increasingly important for electrochemical storage systems include cost, security, independence from foreign competition, and potential markets.

For example, materials based on graphite may become the seat of lithium dendrites during charging. In this context, the demand for the security of users is more and more important when finalizing the battery pack. The use of stable materials and avoiding risk of dysfunction, which could lead to damage, are priorities.

8.5.2. *Development of positive electrode materials*

8.5.2.1. *Current situation*

Compounds with poly-anionic structures have great potential to replace lithium oxides such as $LiCoO_2$.

Among these compounds, the most accomplished material is $LiFePO_4$, which has already been commercialized by Phostech and integrated in an accumulator by several companies such as Valence Technology, A123 Systems and several Chinese companies. This compound, intrinsically an insulator, has been the subject of numerous research studies to change this property. It has a theoretical specific capacity of 170 mAh/g linked to a voltage of 3.45 V versus Li^+/Li. Composite $LiFePO_4/C$ systems have been developed that can reach a reversible specific capacity of 160 mAh/g. The principal interest of these poly-anionic structure compounds resides in their stability, even at high temperature, which makes them suitable for lithium-ion accumulators with large capacity and with reinforced security.

8.5.2.2. *In the medium term*

Among the positive electrode compounds that could replace the cobalt oxide, $LiCoO_2$, and its derivatives, which are currently used in almost all lithium-ion accumulators, high potential insertion materials such as spinel structured oxides (for example: $LiNi_{0.5}Mn_{1.5}O_4$) and olivine structured materials (for example, $LiCoPO_4$) could allow stored energy densities to be improved. New lamellar oxides of the $Li_{1+x}(Mn,M)_{1-x}O_2$ (M = Ni, Co, etc.) type can also give higher capacities when the voltages for a charge cut are sufficiently high (4.5-5 V versus Li^+/Li).

In practice, the use of these positive electrode materials leads to the problem of the reactivity of the electrolyte, which is in contact with an electrode operating at high voltages. This leads to a significant self-discharge effect (up to 80% per month). Yet, the solvents used in the electrolytes for lithium-ion accumulators have an intrinsically high oxidation stability, if we refer to values calculated by *ab initio* methods [ZHA 01] (up to 5.5 V versus Li^+/Li or more for EC, PC, DMC).

NOTE: EC = ethylene carbonate, PC = propylene carbonate, DMC = dimethyl carbonate.

Nevertheless, the potential limit values measured in practice vary a great deal, and are always lower when the electrodes employed for measuring are electrodes based on active materials, as shown by the works of Kanamura *et al.* [KAN 96; KAN 01] on electrolytes based on PC (variation of 5 to 4.2 V versus Li^+/Li). The origin of this noted difference may be several things: the "catalytic" effect of insertion materials, traces of water, and the influence of the lithium salt used. In some cases, the decomposition of the electrolyte is accompanied by the formation of a solid film on the surface of the positive electrode [AUR 02; WUR 05].

Following the example of the works on the negative electrode materials, especially graphite, it would seem of interest to undo this phenomenon, by looking to create a stable solid electrolyte interface (SEI) on the positive electrode by the introduction of additives in the electrolyte. This step, which has not yet become widespread, is currently being followed by Ube Industries in Japan [ABE 06]. It aims to reduce the self-discharge by remedying the interface reactivity problem.

High-voltage electrode materials have been of great interest for several years. Nevertheless, no reliable solution exists at this time. It is likely that a stable electrolyte at high voltage would allow more innovation when it comes to active materials.

In the literature, the proposed solutions to reduce the reactivity at the electrode/electrolyte interface consist of creating a deposit of inorganic compounds that are electrochemically inactive (Al_2O_3, $AlPO_4$, TiO_2, ZrO_2). This passive film could avoid the loss of oxygen from an insertion material for high states of charge (a problem that is usually encountered with lamellar oxides), but would not protect the positive electrode as a whole, i.e. *vis-à-vis* oxidation of the electrolyte. In other words, the protection of electrodes at high potential, using *ex situ* methods cannot alone be an efficient solution.

The protection of the positive electrode should therefore be done *in situ* through use of organic compounds dissolved in a traditional electrolyte (such as $LiPF_6$ in EC/PC/DMC) in order to obtain total protection of the electrode. The oxidation of the additive used should allow the deposit of a passive film that is insoluble in the electrolyte, and this should occur at a lower potential than the potential for oxidation of the electrolyte. The film formed on the surface of the electrode should have a sufficient conductivity of Li^+ ions so that the electrochemical performance (for example speed of charge and discharge, internal resistance) of the accumulator is not limited. The presence of additives should not induce electrochemical instability at the level of the negative electrode, by perturbing, for example, the formation of the passive film at the negative electrode.

Studies in the literature on the protection of the positive electrode have only been carried out on the usual electrodes, for example, $LiCoO_2$. An improvement in the cycling performance of $LiCoO_2$ has been shown to occur by using trimethylphosphite [XU 06], phenyl adamantane [WAN 06], cyclohexylbenzene [HE 07], and biphenyl or triphenyl [MAR 80]. The most studied compound, biphenyl, has an oxidation potential of 4.45 V versus Li^+/Li. The quantity of biphenyl present in the electrolyte should be optimized in order to obtain a thin film, covering but not blocking the transport of Li^+ ions. These additives are used in order to polymerize and to block the accumulator during overcharging.

At the international scale, Sanyo, SDI, and LGC are currently working on developing such materials for high-voltage cathodes.

8.5.3. *Development of anode materials*

Current work on active materials for the negative electrode for a lithium battery are centered on synthesis of compounds using nano-metric insertion (reduction of the diffusion pathway of Li^+) while staying in the right region of size/specific surface which is compatible with a practical maximum electrode density (risk of negating the gains in diffusion length and electronic conductivity associated with the reduced active-material particle size by the addition of supplementary interfaces (that is, the current collector/active material and active material/active material) or by reducing the electrode density (higher specific surface)).

8.5.3.1. *In the short term*

The properties of $Li_4Ti_5O_{12}$ as an active anode material for the lithium power battery are very interesting, especially the potential for oxido-reduction of Ti_4^+/Ti_3^+ which is greater than that for Li^+/Li, namely around 1.55 V/Li^+/Li. This criteria is linked to the weak chemical reactivity of electrolytes, to weak dimensional variations between the reduced/oxidized phases and to the great reversibility of electrochemical phenomena, and allows favorable charge and discharge regimes to be attained.

During the initial insertion of lithium, and then the charge-discharge cycles that follow, the phenomena of dilation or contraction of the mesh are minimal. Indeed, the oxidized phase ($Li_4Ti_5O_{12}$) and the reduced phase ($Li_7Ti_5O_{12}$) have a practically identical volume ($\Delta V = \pm 0.07\%$). For this reason, $Li_4Ti_5O_{12}$ is considered to be a "zero constraint" material. An excellent cohesion at the heart of the electrode and between the electrodes, separators and current collectors is, therefore, maintained throughout the cycling. This avoids the appearance of fissures and avoids damage to the matrix that could lead to a loss of capacity and performance. The theoretical specific capacity is 175 mAh/g.

In practice, the recovered capacity is close to this value for slow cycling regimes. The power performances are more or less high, according to the material synthesis mode and the morphology of the grains. The potential for lithium ion insertion is above the potential for reduction of the electrolyte, which means that almost no passive layer is formed at the electrode/electrolyte interface, as the electrolyte is not degraded. Moreover, this material has great thermal and chemical stability. It can be used for cycling in high power regimes, and/or at low temperature.

In addition, due to its high insertion potential compared to the potential for lithium deposition, there is no risk of internal short-circuit under strong charge currents [NAK 06]. Finally, working at this potential, it is possible to use an aluminum current collector, which is lighter than a copper collector (used for graphite electrodes). Conversely, this relatively high potential (negative electrode) does not allow energy densities to be as high as graphite permits. Currently, $Li_4Ti_5O_{12}$ is a commercialized, low-volume material for lithium-ion accumulators that have high power or rapid charging. Applications such as portable tools, intelligent smart cards, or electric traction are envisaged for this material.

However, the practical capacity obtained with $Li_4Ti_5O_{12}$ is already near the theoretical value (175 mAh/g), which is relatively low compared with that of graphite (330 mAh/g). The margin of progression is therefore limited. Some structural forms of TiO_2 (B or H, notably) [MAR 80; BRO 83; TOU 86], which are less compact than anatase TiO_2 possess the same advantages as $Li_4Ti_5O_{12}$ compared with the graphite carbon that is currently used.

Moreover, they have a theoretical capacity that is clearly superior to the lithium titanium oxide (338 mAh/g for TiO_2 (B) versus 175 mAh/g for $Li_4Ti_5O_{12}$). Thus, for electrodes with low surface capacity (<0.5 mAh/cm^2), the maximum capacity obtained at a very slow regime (C/100) is 260 mAh/g [CHO 07]. Recent studies have in practice reached around 60% of this slow regime capacity (C/10) and very recently some authors have obtained 75% of the theoretical capacity using nano-wires of TiO_2 in regimes of up to 10 C. Granulometry, morphology, specific surface and microstructure seem to be the key parameters for improvement of performances [ZAC 88; NUS 97; KAW 91; ZAK 92; ARM 04; ZUK 05].

8.5.3.2. *In the medium term*

Another approach, equally highly studied over the last few years, consists of looking at the phases of the titanium oxides ($Li_4Ti_5O_{12}$, $Li_2Ti_3O_7$, TiO_2, etc.) [GOV 99], which may be structured in different "nano" forms depending on the synthesis conditions: nano-particles (around 50-100 nm), nano-wires, etc. [CHO 07; KIM 07]. Depending on their morphological characteristics (nano-materials, large specific surface, etc.), the capacity of these materials ($Li_4Ti_5O_{12}$ [AMA 01] or TiO_2 (B) [BRO 06]) to attain the higher charge and discharge regimes makes them the best candidates for making an electrode for an intermediary asymmetric system between the battery and the supercapacitor.

Moreover, for these materials there is the chance for improvement by playing on the morphology and the nano-structure via the original synthesis conditions in order to approach capacities close to that of graphite. In addition, optimization of the active environment at the heart of the electrode composite will play a major role in highlighting the remarkable intrinsic performances of these compounds and reproducing them for thicker electrodes (which have a superior surface capacity or one that is equal 1 mAh/cm^2) and could therefore lead to high energy and power densities.

8.5.3.3. *In the long term*

Numerous studies exist concerning, for example, silicon, tin, or metal alloy nano-particles which could replace the graphite that is currently used. It is foreseeable that the 350 mAh/g stored by graphite used in lithium-ion accumulators can be exceeded, reaching more than 1,000 mAh/g (3,800 mAh/g theoretically). Different materials integrating nanometric silicon in the form of a film or particles, deposited or integrated in a conducting carbon matrix, are well on the way to realizing such performances.

Yet, the complete optimization of the electrodes, and of the overall system of an accumulator with these components integrated, will still need several years of research. Indeed, the expansion of the volume of the Si-Li alloy and the isolation of particles (passivation) are problems that still need to be resolved. Finally, their potential for functioning at around 0.5 $V/Li^+/Li$ remains close to that of graphite, and risks limiting the capacity during rapid charge, despite significant energy densities (problems with an increase in lithium dendrites leading to important safety issues).

8.5.4. *Players in the domain*

Internationally, Toshiba and Enerdel are working on $Li_4Ti_5O_{12}$ materials for the anodes. Sony is working on new anodes based on Sn (Nexelion) with a nano-scale approach of hybridization of several metals: CoSnC and CoSn of nanometric size. The amorphous composite (Sn-Co-C) is associated with a LiNiMnCoO cathode, which allows 90% of the capacity to be charged in 30 minutes at 2 C, with 300 discharge cycles at 1 C (900 mAh capacity, 3.15 Wh of stored energy). Samsung SDI is studying SiO_2-Si and SANYO is working on deposits of metallic particles of silicon linked to a carbon powder, for mid 2010. Panasonic is developing Li-ion rechargeable batteries that use a silicon alloy anode and that are predicted to hit the market in 2012.

8.5.5. *Developments in electrolytes*

8.5.5.1. *High-voltage electrolytes*

The non-aqueous electrolyte is generally a conducting salt such as lithium hexafluorophosphate ($LiPF_6$) dissolved in a mixture of organic solvents that are (cyclic or acyclic) carbonates. It has a conductivity that is clearly less than aqueous electrolytes (around 10 mS/cm versus more than 1 S/cm for sulfuric acid or potassium hydroxide for example), which is in fact compensated by the use of a micro-porous separator of very low thickness, typically less than 20 μm.

The main areas for development of these electrolytes are: i) the extension of the operating temperature range from –20°C to +60°C towards the range of –40°C to +85°C, and ii) the improvement of security for non-inflammable electrolytes.

In addition, the electrolyte is currently the limiting component in the development of high-voltage lithium-ion accumulators (which are unstable at voltages above 4.5 V versus Li^+/Li).

The generic objective of developments regarding high-voltage electrolytes is therefore to limit the reactivity of the positive electrode of the lithium-ion accumulator operating at a high potential, typically greater than 4.5 V versus Li^+/Li. One solution that is envisaged is the formation of a passive conducting film of Li^+ ions on the electro-active surface of the electrode (active material + conducting additives + collector), with the Li^+ ions resulting from the decomposition/polymerization of an additive that was introduced in a conventional electrolyte, following the example of the SEI generated on graphite.

Different types or families of additives can be foreseen (Figure 8.6): polymerizable ones (dioxolane, biphenyl, etc. [ABE 04]), compounds based on phosphorus (trimethyl phosphite [XU 06]), compounds based on boron (LiBOB [CHE 06]), compounds based on sulfur (sulfones, sultones), LiF, and liquid ionic compounds.

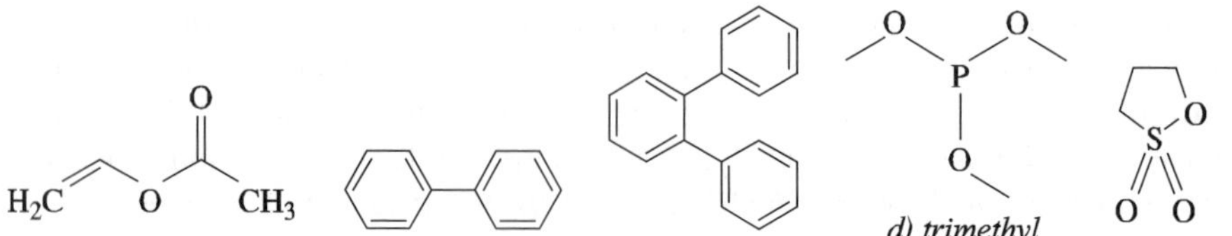

a) Vinyl acetate *b) Biphenyl* *c) o-terphenyl* *d) trimethyl phosphite* *e) γ-sultone*

Figure 8.6. *Foreseeable additive families for high-voltage electrolytes*

8.5.5.2. *Ionic liquids*

Due to the high vapor pressure and the low flammability point of organic solvents, as well as the low thermal stability of $LiPF_6$, the electrolyte decreases the safety of lithium-ion accumulators. The non-volatility, non-flammability, and the good thermal stability of ionic liquids explains the great interest in creating new systems of electrolytes made of these compounds, to replace traditional systems.

Ionic liquids are salts made of organic cations, complexed with inorganic or organic anions, which have the property of being in liquid state around ambient temperature. Ionic liquids are non-volatile (no diffusion in the atmosphere), non-flammable (no risk of explosion), stable at high temperatures (up to 200°C or 400°C depending on the complex), hydrophiles or hydrophobes (depending on the nature of the anion), with viscosity that varies from 37 to 500 cP (at 293 K). They are good conductors (from 0.1 to 15 mS/cm) and they have a wide electrochemical window (between 4 and 5 V). Therefore, they may be used as solvents to replace volatile organic solvents. Indeed, ionic liquids have several advantages that make them attractive: they are better for the environment (non-volatile and non-flammable) and their physic-chemical characteristics can be adapted for a given application by changing the nature of the cation/anion couple. Finally, they have great solvent power, as much for organic as inorganic compounds.

An ionic liquid that is capable of operating as a support electrolyte in rechargeable batteries must have the following properties: i) a large electrochemical window; ii) a weakly coordinating nature for the anion in order to reduce the appearance of ions (a phenomenon that is never totally absent in liquids, taking into account the small size of the Li^+ ion); iii) a low viscosity at ambient temperature in order to increase the mobility of ionic entities; and iv) a wide range of operating temperatures (i.e. high thermal stability and fusion point).

Recent studies have shown that 1,3-dialkylimidazolium salts, although they are very fluid conductors, cannot be used as electrolytes in lithium batteries due to the very positive cathodic potential of 1,3-dialkylimidazolium cations (around 1 V compared to Li^+/Li). Quaternary ammonium (QA) compounds, although very stable electrochemically and easy to synthesize, have until now been less used than imidazoliums as cations in ionic liquids, due to the high fusion points generally observed for obtained compounds. However, ionic liquids prepared using cations that come from quaternary cyclic ammonium compounds, such as pyrrolidinium and piperidinium, have lower fusion points and viscosities, as well as conductivities that are clearly higher than for ionic liquids prepared using acyclic cations. Their properties are comparable to the values observed for ionic liquids based on imidazolium cations (for example, for 1-butyl-1-methylpyrrolidinium TFSI, F=–18°C). Moreover, these properties seem to improve as the cation size diminishes.

It has been shown that ionic liquids based on relatively small saturated cations of the quaternary ammonium type, with an electrochemically stable anion coordinating, such as bis(trifluoromethanesulfonyl)imide ($[(CF3SO2)2N]^-$, $[TFSI]^-$), offer encouraging results. These promising properties include: i) large electrochemical windows due to the low cathodic potential of saturated QA cations and the high anodic potential of $[TFSI]^-$; ii) low viscosities owing to the high flexibility and a good charge distribution in $[TFSI]^-$; and iii) a wide stable liquid domain owing to a low fusion point and a high thermal stability of the $[TFSI]^-$ salt.

Despite these desirable properties, the molar mass and molar volume of the anion $[TFSI]^-$ constitute an evident bottleneck in the improvement of the fluidity and ionic conductivity of the electrolyte. Therefore, this is an immediate barrier for the application of electrolytes based on $[TFSI]^-$ ions in lithium batteries, as a low conductivity cannot supply the demand for high power. In order to increase the properties of ionic liquids based on quaternary ammonium compounds for applications such as electrolytes in systems with high energy density, it is clearly necessary to create stable anions with improved properties, such as lower molecular weights and ion sizes.

In non-aqueous solvent systems, it has been shown that salts of the $Li[C_2F_5BF_3]$ type could offer performances that are comparable to the industrial standard compound, $LiPF_6$, which indicates that anodic and cathodic stability of the $[RFBF_3]$ ion is sufficient for an application such as the electrolyte within lithium batteries. Another method consists of linking the most interesting ionic liquids with a traditional electrolyte that is less viscous.

8.6. Processing and recycling of lithium batteries

Market demands resulting from the electronic energy needs of the general public and the evolution of battery technology, have lead to Ni-Cd batteries, Ni-MH batteries, and, currently, to lithium (lithium ion or lithium polymer) batteries, which are the best performing batteries used today.

Beyond the numerous advantages linked to their performance, lithium systems remain an important source of products, which have a recognized impact on the environment. Their effects are linked mainly to the following components: heavy metals, conducting salts based on fluorine or arsenic (for some cells), organic solvents, and lithium, a very reactive alkaline metal.

Therefore, environmental and regulatory factors justify the future interest of lithium cells and batteries. Recycling the latter is desirable, but it does not seem to be possible in a profitable way for some models that contain expensive compounds (for example, cobalt is sometimes present, but in small quantities). In all cases it is at least necessary to ensure that these accumulators and cells have been rendered inert.

At the end of their life, lithium batteries generate a greater and greater flow of waste, which should be treated. This processing is especially sensitive due to the potential dangers of this kind of waste: lithium-ion batteries must be manipulated with great precaution due to their reactivity.

In France, a company (Recupyl) uses a pilot installation that can specifically process all types of cells and batteries with a lithium anode, using a hydrometallurgic method at ambient temperature. This process allows recycling without pollution and with a maximum rise in value ensuring that the waste is converted into a resource.

Hydrometallurgy covers the set of metal extraction processes, which are achieved by putting the substance in solution (acid or base) and treatments such as lixiviation, electrolysis, and selective precipitation. A hydrometallurgic process comprises the following unitary operations:

– putting the fraction of the substance that contains the chemical element to be developed into solution;

– purification and concentration of solutions being processed;

– transformation to the metallic state using diverse methods that depend on the nature of the metal.

At the scientific and technological level, it seems that the role of hydrometallurgic treatment is fundamental in the elimination and recuperation of heavy metals. Other specialized companies (Citron, Valdi, SNAM) destroy lithium elements using thermal processes by distributing them in small quantities amongst other cells (alkaline and saline) in a pyrolysis oven.

Optimal recycling of the battery would require a design that takes into account the rise in value of its components – an eco-design of the battery elements by the manufacturers and by the chemical suppliers already implicated in recycling. For example, Unicore, world leader in cobalt-based materials required for rechargeable lithium-ion batteries, has today is at the pinnacle of precious metal recycling on the world scale, with a supply of almost 100% of secondary materials by recycling old mobile phones – inclusive of the battery and accessories – and using its own recycling technology.

8.7. Other batteries

8.7.1. *μbatteries, printed batteries, etc.*

Ever since 1991, Japan has had three large-scale research programs on the miniaturization of industrial components, sensors, controllers, etc., and was soon followed in this by the USA and European countries. These projects have led to the emergence of new fabrication techniques, or techniques that have been adapted from previous technologies, especially in the domain of microelectronics. These techniques are often known as micro-techniques, micro-technologies, and even nano-technologies.

Energy sources have undergone the same evolution and micro-generators – micro-cells if they are not rechargeable, micro-batteries if they can be recharged – with a thickness of no more than a few micrometers have begun to appear.

Although they are still only at the experimental stage, these micro-generators have opened the way to numerous applications, notably those using smart cards, micro-machines, electronic tickets, etc. They should not be confused with mini-cells – "paper cells" or polymer electrolyte cells – which deliver much higher currents, but whose thickness is several tens of millimeters.

An *electrochemical micro-generator* is defined as being a system with a total thickness of a few micrometers, prepared by specific thin-layer techniques, cathodic pulverization, thermal evaporation, etc. It is used to supply energy to micro-systems (micro-electronics, micro-machines, etc.).

A micro-generator, that is created using thin layer processes from micro-electronics, should not be confused with other systems using thicker layers (several tens of micrometers for each layer, for example, created using ink impressions based on micronic powders), which are capable of supplying a more significant current for applications in micro-systems. Every elementary cell of a micro-generator has a thickness of the order of 10 μm; it is possible to superimpose them and connect them in series or parallel according to needs. Therefore, it is foreseeable in the short term to have micro-generators on a rigid or flexible support with a surface and thickness that are adaptable depending on the volume available and on the performances required; they will eventually be able to be recharged remotely.

The materials used in micro-generators are generally the same as in lithium batteries. However, the thin-layer fabrication gives them certain specifications: their chemical composition may be far from their larger equivalents, and their structure is generally less organized. Their electrochemical properties can therefore be quite different from larger materials. Thus, starting with known bulk materials, using thin-layer technology, we can obtain new materials with new properties that have to be characterized.

Metallic lithium is the main *negative* electrode material studied today in micro-generators due to its numerous advantages. The behavior of this material is well characterized in traditional batteries. Its development in thin-layer form does not seem to create any particular problems. Yet, certain laboratories are researching alternative solutions, for lithium is not without its disadvantages. First of all, it is deposited using thermal evaporation, whereas the electrolyte and the positive electrode are deposited by other methods. A single deposition technology would be preferable. In addition, lithium is very sensitive to humidity; therefore, protection is necessary, without increasing the resulting thickness too much. To compensate for these disadvantages, it is possible to replace lithium by an insertion material, similar in principle to that which is used at the positive electrode, but with as low a potential as possible. In this way, a micro-generator with two insertion electrodes is realized, known as the *rocking-chair* micro-generator. Depending on the targeted application, the loss of potential difference, compared to a lithium micro-generator, is not necessarily harmful. Nevertheless, attempts to manufacture the negative electrode with insertion materials are rare. We can point out Nb_2O_5 and $Li_4Fe_{0.5}Ti_{4.5}O_{11.75}$ which are used in a completely solid micro-generator. The solution of replacing lithium with carbon, which is used in larger batteries (lithium-ion batteries), does not seem to be transferrable to micro-generators. The search for a negative electrode using an insertion material other than carbon is therefore a problem that is specific to micro-generators, which implies that there is still a lot to be done in this field.

The ideal positive electrode material is a mixed electronic and ionic conductor, capable of being reversibly inserted around a lithium ion by a transition element. The potential corresponding to the oxido-reduction reaction of the transition element must be as high as possible, the molar mass must be as low as possible, in order that the specific energy can be as high as possible.

Titanium disulfite, TiS_2, was the first positive electrode material for lithium batteries. Completely solid micro-generators using this material were manufactured at the start of the 1980s. Thus, TiS_2/glass/Li systems have been able to be cycled 1,000 times, between 2.6 and 1.4 V, at current densities of 0.1 to 0.5 mA/cm^2. The obtained capacity, 50 μAh/cm^2, corresponds to the reversible insertion of 0.9 lithium ions by titanium, and it is constant throughout the cycles. The surface of such systems varies from 1 to 10 cm^2 and should be able to be increased. These performances, sufficient for some applications, would allow the industrial development of micro-generators. However, micro-cells and micro-generators are still at the research stage; the most advanced manufacturers are today delivering prototypes and are working actively to start production.

In theory, transition element oxides are better positive electrode materials than their sulfurated equivalents. Moreover, they are now exclusively used in lithium batteries. For several years there have therefore been numerous attempts to prepare them in thin-layer form and to use them in micro-generators. As a general rule, thermal processing is required after deposition for these materials. The aim is to make the diffusion of lithium easier by improving the crystallinity of the thin layers. This thermal processing may be problematic if, for example, the micro-generator is to be placed on a chip.

One of the cathode materials that has been studied the most is $LiMn_2O_4$, which is prepared using cathodic pulverization or thermal evaporation. A $LiMn_2O_4$ /*Lipon* /Li micro-generator has been cycled 300 times, at 10 μA/cm^2 between 4.5 and 3.8 V. Its almost constant capacity is 15 μAh/cm^2 and is only reduced from 40% to 100 μA/cm^2. This value of 15 μAh/cm^2 is relatively low compared to the value obtained with TiS_2, and this is due to the low thickness of the thin layers of $LiMn_2O_4$. Other oxides present similar performances, such as $LiNiO_2$, $LiCoO_2$, and V_2O_5.

8.7.2. *Electrolytes*

Electrolytes must be stable with respect to lithium, and have a domain of electrochemical stability of several volts. In the thin-layer form with a thickness of around 1 μm, some solid electrolytes have a sufficiently low ionic resistance, of the order of 100 Ω (from around 10^{-7} to 10^{-4} S/cm).

The main electrolytes are inorganic materials, generally glasses based on oxides which are prepared in thin-layer form using cathodic pulverization. For example, vitreous thin layers with the $(B_2O_3)_{0.38}(Li_2O)_{0.31}(Li_2SO_4)_{0.31}$ composition which have an ionic conductivity of 10^{-7} S/cm at ambient temperature. This value is sufficient to obtain currents of 100 μA/cm^2 without excessive polarization. However, a more significant conductivity is necessary for micro-generators to function at a higher current or at a lower temperature. One of the strategies to obtain that result is to deposit thin layers using reactive cathodic pulverization, under partial pressure of nitrogen. Thus, the glass, $Li_3PO_{2.5}N_{0.3}$, known as *Lipon*, which is obtained by this process using a target of $(P_2O_5)_{0.25}(Li_2O)_{0.75}$, has an ionic conductivity of 3×10^{-6} S/cm at ambient temperature. This glass is very stable when in contact with lithium, and its domain of electrochemical stability is of the order of 5 V.

It is equally possible to incorporate sulfur into the glass oxides. A glass from the P_2O_5-P_2S_5-Li_2O-LiI system has an ionic conductivity of 5×10^{-5} S/cm at ambient temperature. However, as it is very reactive with lithium, a very fine layer of LiI is used to insulate it, even though the ionic conductivity of the glass/LiI combination is reduced to 2×10^{-6} S/cm at ambient temperature. The domain of electrochemical stability of this combination is around 3 V.

Ionic conductivities of 10^{-4} S/cm were obtained for thin layers of glass sulfites from the GeS_2-Ga_2S_3-Li_2S system. However, these materials are as hygroscopic as lithium and are, therefore, difficult to manipulate.

8.7.3. *Rocking-chair micro-generator*

An example of the rocking-chair micro-generator is a micro-battery made up of a negative electrode with lithium inserted, with the composition of $Li_4Fe_{0.5}Ti_{4.5}O_{11.75}$, an electrolyte with the composition of $LiBO_2$ (a low performing electrolyte) and a positive electrode with the composition of $LiCoO_2$. A very fine layer of $LiTaO_3$ is inserted between the electrolyte and the negative electrode in order to prevent the formation of a passivation layer. This system can be cycled between 2.6 and 1 V, at 100 μA/cm^2. The capacity is around 35 μAh/cm^2.

8.7.4. *Manufacturing techniques*

Only the chemical vapor phase deposit (CVD) or physical vapor phase deposit (PVD) techniques can obtain low thickness layers with the appropriate properties (Figure 8.7).

Layers obtained by CVD are usually crystallized; they have good adherence and the deposition speed can be increased. As the substrate is heated to several hundred degrees Celsius, this technique does not allow the use of flexible and thin materials, such as plastics, which must be used in order for the micro-generator and its support to have a low thickness. It does not seem as though this technique will be developed despite good performance of the resulting batteries. Only the positive (TiS_2) is manufactured using CVD.

The thin layers obtained by PVD require thermal evaporation and cathodic pulverization. Thermal evaporation is not the best adapted process for obtaining electrode or electrolyte material. Numerous compound substances will be partially or totally decomposed with a loss of the most volatile elements. The energy of the evaporated particles is of the order of a few tens of electronvolts, which leads to a low adherence on the substrate and does not lend itself to the creation of multiple layers.

Conversely, *cathodic pulverization* is the method of choice for these types of applications. The composition of the deposit is often very close to that of the target. Compounds, such as oxides and sulfides, are usually obtained in amorphous form at ambient temperature, which is an important advantage for the positive electrode. The insertion of lithium is done as an isotrope and without much dimensional modification, due to the low density of the deposited material. This advantage allows for a large number of charge-discharge cycles without harming the micro-generator. The energy of the particles, at a few electronvolts, leads to good adherence of the layers, with the occasional disadvantage of the appearance of internal tensions, which can cause tears.

As many materials are insulators or bad electricity conductors, radiofrequency cathodic pulverization will be used with the addition of the magnetron effect, in order to obtain significant deposition speeds and to reduce the pulverization pressure. The quality of the layers will therefore be improved; notably the porosity will be much lower. Recall that the electrolyte film should be totally free of holes in order to avoid short-circuits.

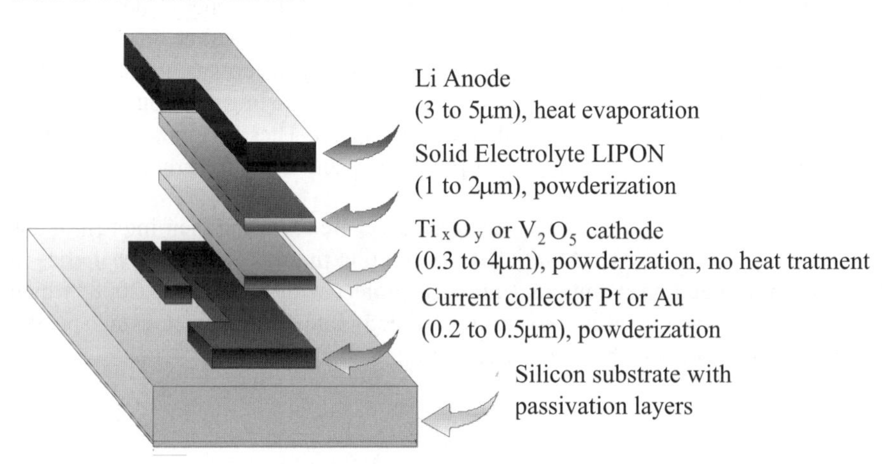

Figure 8.7. *Fabrication techniques of a lithium micro-battery*

8.7.5. *Printed batteries*

At least 1 mm thick, ultra-thin batteries are ideal for intelligent tickets, for radio frequency identification tickets (RFID), for autonomous sensors, and for active packaging applications requiring an external energy source. They can serve to replace button cell batteries in applications as diverse as greeting cards, purchase displays, and even customer loyalty cards. The strategy for development of ultra-thin batteries uses high-speed impression technologies to make an accessible product with a low cost, which is thinner and more flexible.

Currently, the majority of ultra-thin battery products use primary (non-rechargeable) elements such as MnO2/Zn (0.5 mm thick) with capacities of 10 to 30 mAh and lifespans of around 3 years. Blue Spark, for example, produces 1.5 V (around 12 mAh) printed batteries based on carbon-zinc, which deliver current peaks greater than 1 mA, for a thickness of 500 µm. Further upstream, NEC is developing an organic (rechargeable) radical battery, but this system only delivers very little energy and is limited to capacity type power applications (around 0.1 mAh/cm^2).

Finally, research and development work is being carried out on nano-batteries (some using the properties of carbon nano-tubes) and also bio-batteries (Sony).

8.7.5.1. *Lithium-ion printed batteries*

The growth in demand for autonomous energy sources is more and more significant for applications that need a variety of innovative lithium-ion accumulator architectures. Lithium-ion technology offers the best specific energy densities and lithium-ion accumulators are overtaking the mobile energy market. In order to be able to create certain kinds of batteries for which the usual formats are not suitable, one approach consists of printing the electrodes using existing printing techniques, in order to create electrode patterns that are in demand. These printing technologies limit material losses, increase the rate of production, and make the process flexible.

Today, few of the known battery manufacturers are revealing their interest in this innovative approach. The CEA initiated this printing process for lithium batteries in 2004 and has begun an ambitious European project with VARTA in 2008, which aims to print a complete lithium-ion battery directly into a host object, using specific patterns that can only be obtained using printing technologies.

8.8. Bibliography

[ABE 04] ABE K. *et al.*, "Functional electrolyte: additives for improving the cycleability of cathode materials", *Electrochem. Solid-State Lett.*, vol. 7, p. A462-A465, 2004.

[ABE 06] ABE K., USHIGOE Y., YOSHITAKE H., YOSHIO M., "Functional electrolytes: Novel type additives for cathode materials, providing high cycleability performance", *J. Power Sources*, vol. 153, p. 328, 2006.

[AMA 01] AMATUCCI G.G., BADWAY F., DU PASQUIER A., ZHENG T., "An asymmetric hybrid nonaqueous energy storage cell", *J. Electrochem. Soc.,* vol. 148, pp. A930–A939, 2001.

[ARM 04] ARMSTRONG A.R. *et al.*, "TiOz-B nanowires", *Chem. Int. Ed.* vol. 43, pp. 2286-2288, 2004.

[AUR 02] AURBACH D. *et al.*, "On the capacity fading of $LiCoO_2$ Intercalculation electrodes: the effect of cycling, storage, temperature and surface film forming additives", *Electrochimica Acta*, vol. 47, p. 4291, 2002.

[BOS 06] VAN DEN BOSSCHE P. *et al.*, "An assessment of sustainable battery technology", *J. Power Sources*, vol. 162, pp. 913-919, 2006.

[BRO 83] BROHAN L., MARCHAND R., "Properties physiques des bronzes $MxTiO_2$(B)", *Solid State Ionics*, vol. 9-10, pp. 419-424, 1983.

[BRO 06] BROUSSE T., MARCHAND R., TABERNA P.-L., SIMON P., "TiO2(B)/activated carbon non-aqueous hybrid system for energy storage", *J. Power Sources*, vol. 158, pp. 571-577, 2006.

[CHE 06] CHEN Z., LU W.Q., CHOW T.R., AMINE K., "$LiPF_6$/LiBOB blend salt electrolyte for high-power lithium-ion batteries", *Electrochim. Acta*, vol. 51, p. 3322-3326, 2006.

[CHO 07] CHO K., CHO J., "Rate characteristics of anatase TiO_2 nanotubes and nanorods for lithium battery anode materials at room temperature", *J. Electrochem. Soc.*, vol. 154, pp. A542-A546, 2007.

[GOV 99] GOVER R.K.B. *et al.*, "Investigation of ramsdellite titanates as possible new negative electrode materials for Li batteries", *J. Electrochem. Soc.* vol. 146, no. 12, pp. 4348-4353, 1999.

[HE 07] HE Y-B. *et al.*, "The cooperative effect of tri(β-chloromethyl) phosphate and cyclohexyl benzene on lithium ion batteries [J]", *Electrochim. Acta*, 52, p. 3534-3540, 2007.

[JOS 05] JOSEFOWITZ W. *et al.*, "Assessment and testing of advanced energy storage systems for propulsion", in *European Testing Report: the 21st International Battery, Hybrid and Fuel Cell Electric Vehicle Symposium & Exposition*, Monaco, April 2, 2005.

[KAN 01] KANAMURA K. *et al.*, "Oxidation of propylene carbonate containing $LiBF_4$ or $LiPF_6$ on $LiCoO_2$ thin film electrode for lithium batteries", *Electrochimica Acta*, vol. 47, p. 433-439, 2001.

[KAN 96] KANAMURA K. *et al.*, "Studies on electrochemical oxidation of non-aqueous electrolyte on the $LiCoO_2$ thin film electrode", *J. Electroanal. Chem*, vol. 419, issue 1, p. 77-84, 1996.

[KAW 91] KAWAMURA H., MURANISHI Y., MIURA T., KISHI T., "Lithium insertion characteristics into titanium oxide", *Denki Kagaku oyobi Kogyo Butsuri Kagaku*, vol. 59, pp. 766-772, 1991.

[KIM 07] KIM J., CHO J., "Spinel $Li_4Ti_5O_{12}$ nanowires for high-rate Li-ion intercalation electrode, electrochemical and solid-state letters", *E.S.S.L.*, vol. 10, pp. A81-A84, 2007.

[MAR 80] MARCHAND R., BROHAN L., TOURNOUX M., "TiO_2(B) A new form of titanium dioxide from potassium octatatinate $K_2Ti_8O_{17}$", *Mat. Res. Bull.*, vol. 15, pp. 1129-1133, 1980.

[NAK 06] NAKAHARA K., NAKAJIMA R., MATSUSHIMA T., MAJIMA H., "Preparation of particulate $Li_4Ti_5O_{12}$ having excellent characteristics as an electrode active material for power storage cells", *J. Power Sources*, vol. 117, pp. 131-136, 2006.

[NUS 97] NUSPL G., YOSHIZAWA K., YAMABE T., *J. Mater. Chem.*, vol. 7, pp. 2529-2536, 1997.

[RAP 02] RAPPORT DE L'AIE, Use of photovoltaic power systems in stand-alone and island applications. Task 3: Management of storage batteries used in stand alone photovoltaic power systems, 2002, AIE (Agence internationale de l'énergie, Paris, France).

[THE 02] THEMATIC NETWORK, Contract N° ENK5-CT-2000-20336, Project funded by the European Community under the 5th Framework Programme, 1998-2002.

[TOU 86] TOURNOUX M., MARCHAND R., BROHAN L., "Layered $K_2Ti_4O_9$ and the open metastables TiO_2(B) structure", *Prog. Solid State Chem.*, vol. 17, pp. 33-52, 1986.

[WAN 06] WANATABE Y. *et al.*, "Organic compounds with heteroatoms as overcharge protection additives for lithium cells", *J. Power Sources*, vol. 160, p. 1375-1380, 2006.

[WUR 05] WÜRZI A. *et al.*, "Film formation at positive electrodes in Lithium-ion batteries", *Electrochem. Solid-State Lett.*, vol. 8, p. A34-A37, 2005.

[XU 06] XU H.Y. *et al.*, "Electrolyte additive trimethyl phosphate for improving electrochemical performance and thermal stability of $LiCoO_2$ cathode", *Electrochim. Acta*, 52, p. 636-642, 2006.

[ZAC 88] Zachau-CHRISTIANSEN B., WEST K., JACOBSEN T., ATLUNG S., "Lithium insertion in different TiO_2 modifications", *Solid State Ionics*, vol. 28-30, pp. 1176-1182, 1988.

[ZAC 92] ZACHAU-CHRISTIANSEN B., WEST K., JACOBSEN T., SKAARUP S., "Lithium insertion in isomorphous MO_2 structures", *Solid State Ionics*, vol. 53-56, pp. 364-369, 1992.

[ZHA 01] ZHANG X., PUGH J. K., ROSS P. N., "Computation of thermodynamic oxidation potentials of organic solvents using density functional theory", *J. Electrochem. Soc.*, vol. 148, pp. E183-E188, 2001.

[ZHA 06] ZHANG S.S., "A review on electrolyte additives for lithium-ion batteries", *J. Power Sources*, vol. 162, p. 1379-1394, 2006.

[ZUK 05] ZUKALOVĂ M. *et al.*, "Lithium Storage in TiO_2(B)", *Chem. Mater.*, vol. 17, pp. 1248-1255, 2005.

Chapter 9

Supercapacitors: Principles, Sizing, Power Interfaces and Applications

9.1. Introduction

Supercapacitors are components that are dedicated to energy storage, and appeared at the start of the 2000s. They offer significant energy densities and power densities for medium to high-power applications.

Supercapacitors, otherwise known as electric double-layer capacitors, are in fact intermediate components between the conventional capacitor and the battery of accumulators. They are characterized by having a lower energy density than batteries but a greater energy density than capacitors.

The properties of supercapacitors depend on the principles behind their manufacture. The effective surface of the electrodes is considerably increased by the use of powdered charcoal. The size of the equivalent dielectric for a capacitor is defined by the dimensions of the ions at the interface of the two electrodes with the electrolyte ensuring the mobility of ions from one electrode to another. This size is between 2×10^{-10} and 10×10^{-10} m.

Proposed capacitances range from a few farads to a few thousand farads. However, the choice of electrolytes that offer the required ionic conductivities leads to maximum admissible voltages that do not exceed 3 V.

Chapter written by Philippe BARRADE.

At a high capacitance and a low operating voltage, the principal characteristic is the existence of theoretically reversible phenomena during the charge/discharge phases of the component, without any chemical reaction. As a result, the lifespan of these supercapacitors, or the number of cycles possible, is considerable compared with electrochemical accumulators (from 10^5 cycles to 10^6 cycles for supercapacitors).

However, even if the energy densities and power densities of supercapacitors are substantial, a single cell is not enough for the majority of applications. The energy density is limited, and it is necessary to research the number of cells that must be used in a bank of supercapacitors in order to supply the energy needs of a particular application. Following the example of batteries of accumulators, we also talk of batteries of supercapacitors.

In addition, the notion of efficiency during the charge/discharge phases should be taken into account when sizing a bank of supercapacitors, by qualifying the power density of the cells. The notion of efficiency may be more important than the amount of energy to be stored when efficiency defines the number of cells required. The bank of supercapacitors can then be oversized in terms of stored energy.

Finally, the link between a bank of supercapacitors and its application cannot be made directly, as the voltage at the terminals of the cells varies according to their state of charge. A continuous-continuous or continuous-alternating static converter is necessary to adapt the voltage levels, while controlling the charge and discharge currents of the cells. The very low voltage of supercapacitors has an effect on the structure of the converter, which must generally be a voltage elevator.

The search for a converter with elevated efficiency also affects the arrangement of the cells in the bank of supercapacitors. The series arrangement of cells required is that which, for a given power, limits the charge and discharge currents of the accumulator. There are recurrent problems linked to placing very low voltage sources in series, for which solutions involving voltage balancing should be defined.

The energy density of supercapacitors does not allow them to be used as the main source of energy for most applications.

However, their power density and significant lifespan make supercapacitors ideal for any hybrid system application. In a large number of applications, the supercapacitor is used as a buffer reservoir to limit fluctuations in power of the main energy source (network, battery of accumulators, fuel cells, internal combustion engine, etc.).

9.2. Supercapacitor: electric double-layer capacitor

9.2.1. *Principles*

In order to define the supercapacitor and to characterize its properties and its fundamental parameters, its principles of operation are given in Figure 9.1.

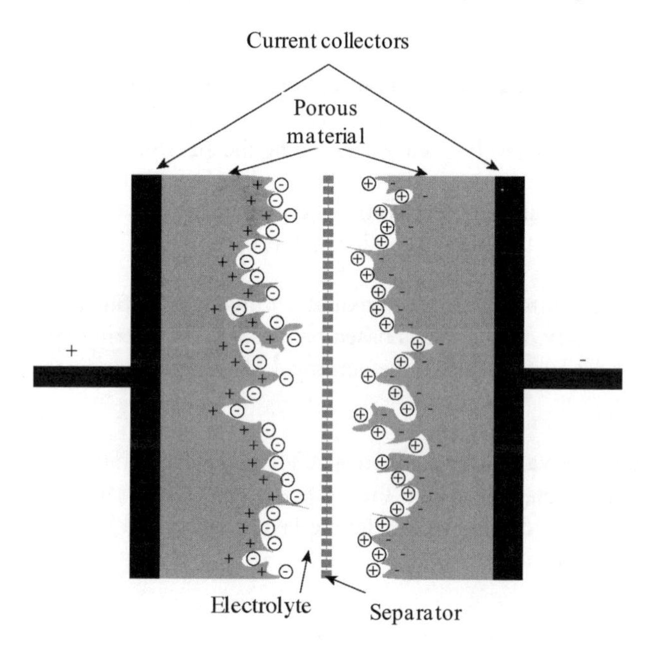

Figure 9.1. *Principles of operation for a supercapacitor*

A supercapacitor consists of two electrodes that are galvanically isolated by a separator, which is soaked in an electrolyte.

The two electrodes are created by deposition of a porous material on a metallic film. The metallic film is generally aluminum, whereas charcoal (activated charcoal) is chosen as the porous material. When the component is charged, the charges are stored at the interface between the porous material and the electrolyte. The use of activated charcoal enables charge to be stored on a significant active surface while offering a good electric conductivity.

The function of the electrolyte is to ensure the mobility of the ions it contains towards the electrodes. The anions should be able to progress freely towards the positive electrode, whereas the cations should be able to progress freely towards the negative electrode. The electrolyte may be solid, but is generally liquid. The choice of the electrolyte is the result of a compromise between the voltage performance and the ionic conductivity. Minimization of the latter leads to a choice of electrolytes that have low dissociation voltages (1 V). In order to avoid oxido-reduction mechanisms, which would lead to irreversible mechanisms during the charge and discharge phases, the operating voltages of the supercapacitor must be limited (from 2.5 to 3 V).

The separator is generally a sheet of paper. Its role is that of an insulator, which should prevent any galvanic contact between the electrodes. However, it must be able to be soaked in electrolyte without reducing the electrolyte's ionic conductivity.

Two principal parameters lead to the energy density of a current supercapacitor: the maximum voltage that can be applied and its capacitance.

As previously introduced, the maximum voltage is defined by the choice of electrolyte. Currenlty, the choice is based on electrolytes offering operating voltages between 2.5 and 3 V, the latter at the detriment of the ionic conductivity, which will be reduced.

The capacitance values for a supercapacitor go from a farad to a kilofarad. Such capacitances are obtained using the principle on which the operation of the supercapacitor is based: the electric double-layer, introduced by Helmholtz in 1879 and shown in Figure 9.2.

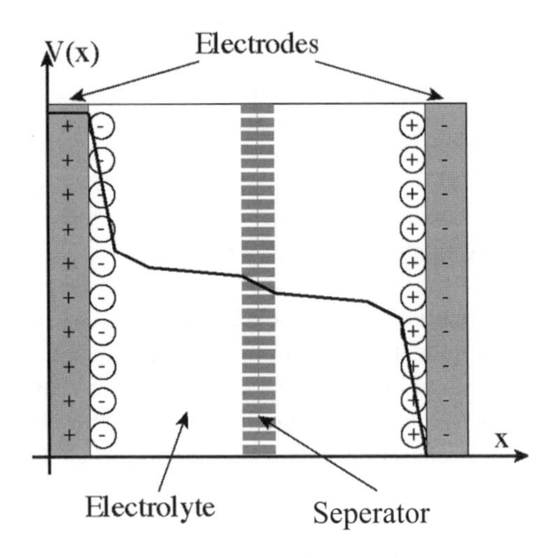

Figure 9.2. *Electric double-layer capacitor*

During charging, anions contained in the electrolyte are attracted towards the positive electrode, whereas cations converge towards the negatively charged electrode. At the interface between each of the two electrodes and the electrolyte, two double-layers are formed, i.e. an accumulation of charges that are different depending on the layer: positive electrostatic charges and anions on the positive electrode, and negative electrostatic charges and cations on the negative electrode. These two double-layers lead to capacitances defined by the following relation:

$$C_{dc} = \varepsilon \frac{A}{d} \qquad [9.1]$$

where C_{dc} is the capacitance of a double-layer (on the positive or negative electrode side), ε is a dielectric constant, *A* is the effective surface of the electrodes, and *d* is the equivalent dielectric value for a traditional capacitor.

The first factor that justifies the capacitances proposed by supercapacitor technology is the porosity of the activated charcoal used in the electrodes, as the effective surface *A*, which is offered for charge storage is considerable (3,000 m^2/g). Conversely, the equivalent dielectric value *d*, is defined by the dimensions of the anions for the positive electrode and by the dimensions of the cations for the negative electrode. In fact, *d* varies between 2×10^{-10} m and 10×10^{-10} m. According to equation [9.1], the capacitances associated to the two double-layers are inversely proportional to *d*, which further increases their values.

Therefore, we can regard the supercapacitor as two capacitors in series with each other, representing the two double-layers of the two electrodes. The equivalent capacitance obtained is due to the effective surface of the two electrodes (which is increased owing to the porosity of the chosen material), as well as the dimensions of the anions and cations. This leads to significant capacitances, of up to a kilofarad. Finally, we must remember that the maximum voltage admissible is determined by the choice of electrolyte. This voltage is between 2 and 3 V.

9.2.2. *Electric model – principal parameters*

Two types of models are generally proposed for supercapacitors: modeling using the equivalent electric diagram, or else modeling using impedance spectroscopy. Modeling using the equivalent electric diagram has the advantage of directly linking its parameters to the component as defined previously. Moreover, it allows the direct linking of the modeling work with the operation of the supercapacitor, which we will explain later (sizing of a bank of supercapacitors, analysis of the efficiency).

Figure 9.3 gives the equivalent electric model of a supercapacitor. This is the most commonly used model [ZUB 00].

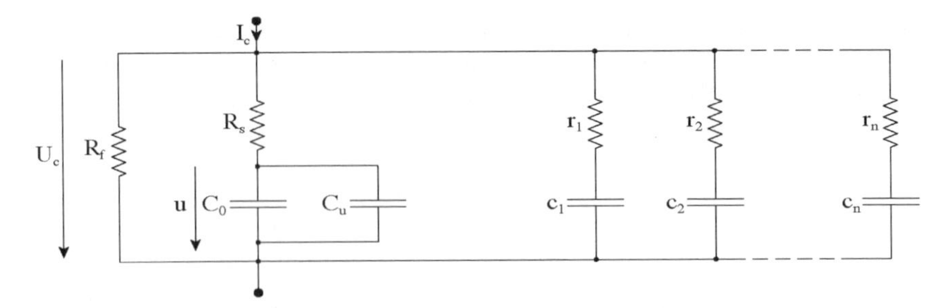

Figure 9.3. *Electric model of a supercapacitor*

The parameter that mainly characterizes the supercapacitor is its capacitance C_0. This value is defined by placing the capacitances of the two double-layers, as defined in equation [9.1], in series with each other. This parameter is given by the manufacturer.

However, measuring of the capacitance of a supercapacitor shows that its value is not constant but is a function of the voltage, u, at the terminals of the two double-layers, and is, therefore, a function of the voltage, U_c, at the terminals of the component. The dependence of the capacitance on the voltage of a supercapacitor is linked to the existence of two diffusion layers within the electrolyte, each directly in contact with each of the two double-layers. These diffusion layers are defined by their size, their respective densities in anions and cations, and the temperature of the electrolyte. Moreover, the potential difference at the terminals of the double-layer, which is associated with the diffusion layer, directly influences the size of the latter. Each of these diffusion layers has a capacitance, whose value is inversely proportional to their size. When the voltage at the terminals of the supercapacitor increases, the size of the diffusion layers is reduced, which leads to an increase in its capacitance. This phenomenon is accounted for in the electric model by the capacitance, C_u, which is defined as a function of the voltage u.

In fact, the capacitance, C, of the supercapacitor is defined by the relation:

$$C = C_0 + C_u \text{ with } C_u = K.u \qquad [9.2]$$

where C_0 is the capacitance constant defined by the two double-layers, u is the voltage at the terminals of the two double-layers, and K is the constant of variation of the capacitance C_u linked to the diffusion layers. For some supercapacitors, the capacitance C_u may have a value of up to 25% of the principal capacitance C_0 under the maximum voltage of the component (2.5 V).

The resistance R_s defines the series resistance of the supercapacitor. Its value is partly fixed by the quality of the deposits of porous material on the metallic sheets to create the electrodes. However, its value is mainly defined by the ionic conductivity of the electrolyte. In fact, the analysis of the electric potential in the supercapacitor, as presented in Figure 9.2, shows a voltage drop in the electrolyte during the charge and discharge phases for the component (charge phase is shown in Figure 9.2). When the component is at rest, this voltage drop in the electrolyte should be zero, provided we do not consider phenomena linked to the diffusion layers. The value of R_s is typically between 0.5 and 100 mΩ. It is provided by the manufacturers.

The resistance R_f defines the leakage current of the supercapacitor, which is larger than the leakage currents characteristic of batteries of accumulators. The resistance of the leak is principally linked to the electric conductivity of the separator, as well as to the impurities contained in the electrolyte. Its value decreases when the cell is charged beyond the maximum voltage possible (dissociate voltage of the electrolyte, phenomena of oxido-reduction at the interface between the porous material and the electrolyte). The value of R_f is typically between 500 Ω and 100 kΩ. Its value is not systematically provided by manufacturers, who instead cite the leakage current under the maximum voltage, which is between 40 μA and 10 mA.

Other than the main supercapacitor parameters, which have just been defined, the model given in Figure 9.3 adds a set of RC cells in parallel (r_1c_1, …, r_nc_n). These cells characterize phenomena of charge redistribution, or relaxation, characterized by time constants, which are between a few seconds and several hours, or even more. These charge redistributions correspond to the movement of stored charges from the easy access zones of electrodes towards limited access zones. This is because of the great porosity of activated charcoal. Therefore, on the electrodes we see a non-uniform distribution of the charges during the rapid charge of a supercapacitor. Once the charging is finished, free circulation of charges on the electrodes, with multiple time constants, finally leads to a homogenous distribution.

The behavior at charging and discharging is presented in Figure 9.4. For a given charge/discharge current, we give the evolution of the voltage, U_c, at the terminals of a supercapacitor with 1,500 F/2.7 V, for which the model presented here adheres to the behavior of the component.

The charge and discharge of supercapacitors are carried out at constant current. Figure 9.4a shows the dependence of the capacitance of the components on the voltage at its terminals. Compared to the vector (1), the voltage does not evolve linearly. Moreover, the segment denoted (2) illustrates the presence of the series resistance R_s by a sharp fall in the voltage U_c when the charge current is stopped.

Figure 9.4b shows a complete charge/discharge cycle. After charging of the component, and the voltage drop linked to the series resistance, the voltage at the terminals of the component decreases significantly, not because of the leakage current, but due to the relaxation phenomena: the natural redistribution of charges on the two electrodes to reach a uniform distribution leads to a voltage drop that is not linked to losses. There is a dual phenomenon after discharge of the component: the relaxation phenomenon leads to an increase in the voltage even though the discharge current is zero.

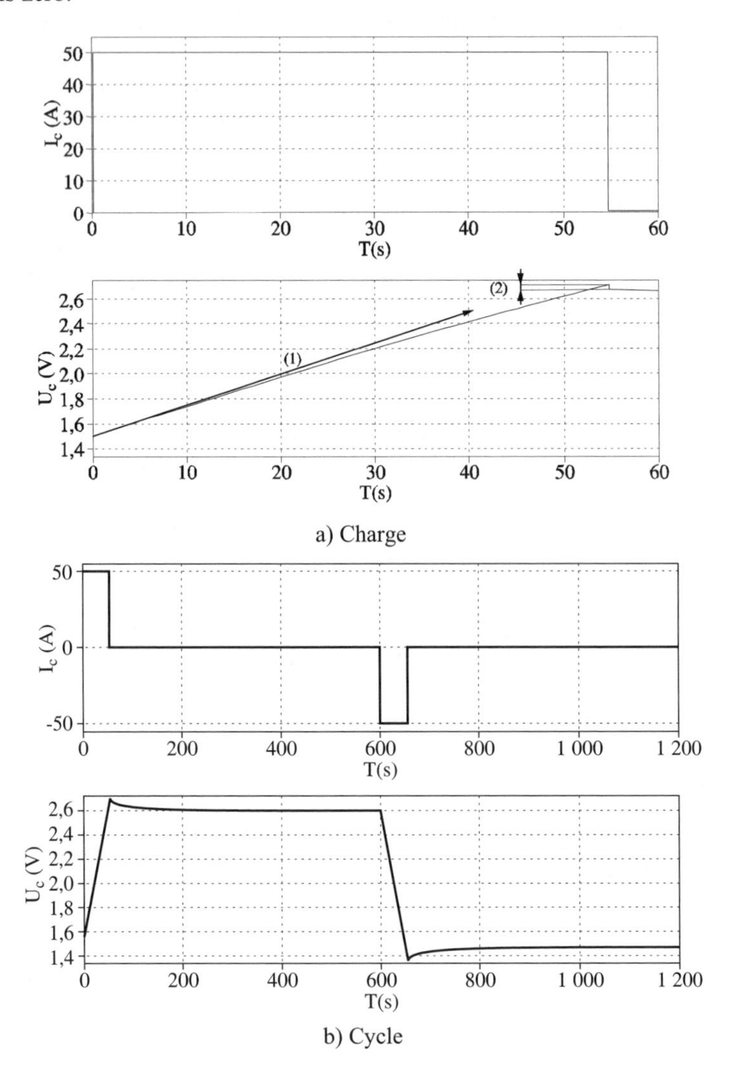

Figure 9.4. *Charge and discharge of a supercapacitor*

The values of the parameters linked to relaxation phenomena are not usually given by the manufacturers, although they are generally favorable: whether the charging of the supercapacitor can be done in several minutes or several hours, the capacitances associated with relaxation allow more energy to be stored than that that could be predicted by only considering the principal capacitance. However, it is difficult to exploit these extra capacitances, especially when the supercapacitor is cycled with time constants of the order of a minute; consequently, we cannot evaluate the energy associated with the relaxation capacitors.

In fact, using the model given in Figure 9.3 generally leads to the RC relaxation circuits not being used.

9.2.3. *Thermal model*

As explained in the presentation of the equivalent electric model, the supercapacitor contains a series resistance, which, although of low value, leads to losses during charge/discharge phases, and therefore, leads to a dissipation of energy due to the Joule effect. Therefore, the supercapacitor heats up, which should be characterized in order to determine whether the temperature reached may be tolerated, or whether the cooling system should be utilized.

The definition of a precise thermal model is difficult as it requires precise knowledge of the internal structure of the component, which is not generally given by manufacturers. Manufacturers propose tools that enable the heating of the device to be calculated in terms of the requests. However, these tools are closed, and are the result of a model, which is not given. Therefore, in the majority of cases, we must be satisfied with a "black box" model, using a few parameters given directly by the manufacturers and estimating those that are not given. Among the given parameters are the thermal conductivity of the component and its density (found from its volume and weight). The thermal capacitance is not given. It can be found either by appropriate measurements or by estimation.

Indeed, to the extent that the nature and constitution of the components of the supercapacitor are known, we can establish the thermal capacitance of the device from the thermal capacitances of each of the constituents. For example, for a device that consists of 65% aluminum (900 J/kg/K), 25% carbon (900 J/kg/K) and 10% electrolyte (4,180 J/kg/K), we obtain a global thermal capacitance of 1,203 J/kg/K.

Thus, using software for finite element numerical analysis, the thermal conductivity, the component density, and the thermal capacitance enable the thermal behavior of the devices being used to be determined. An example of a simulation in two dimensions is given in Figure 9.5.

Three supercapacitors are mounted on a printed circuit board (PCB), within a carbon fiber box that is open at its two sides. The top of the box is subjected to the sun's rays (1 kW/m^2), a flow of air at constant speed circulates from the left to the right (0.5 m/s), while the losses due to the charge/discharge currents in the series resistances come to 1.13 W in each of the three supercapacitors, during a cycle of 2,400 s.

The proposed result shows the heating of the three devices. Partial information is provided by measuring the temperature on the walls of the devices, as the internal temperature is higher.

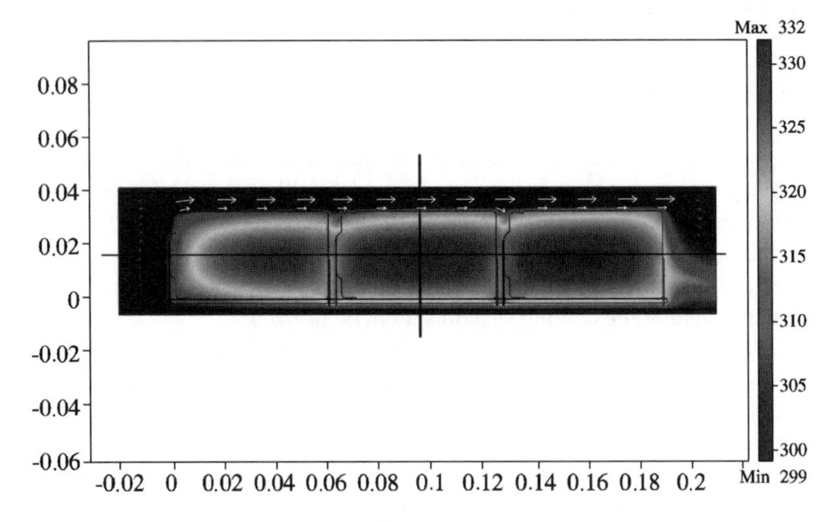

Figure 9.5. *Thermal model of a supercapacitor*

We should note that such a model is nevertheless a global model: the losses in the components are considered to be uniformly distributed, and the devices are considered to be homogenous. This is not strictly the case, but such a model remains close to the real behavior of the component.

9.3. Sizing a bank of supercapacitors

9.3.1. *Energy criteria*

For most applications, a single device is not sufficient to ensure that energy needs are met; therefore, it is necessary to determine the number of devices required.

For this, we first consider the electric model defined in Figure 9.3, neglecting the parameters linked to relaxation and considering the principal constant capacitance C ($C=C_o$). In this case, the maximum energy, W_M, that can be stored in a device under the maximum voltage admissible, U_M, is defined by the following relation:

$$W_M = \frac{1}{2} C U_M^2 \quad [9.3]$$

Using the maximum voltage U_M, we have to decrease the voltage towards 0 if we wish to extract the energy stored in the device. However, at a given power, the current in the supercapacitor should tend towards infinity when the voltage tends towards zero. This cannot be tolerated for questions of efficiency [BAR 03a], [BAR 03b]: losses in the series resistance of the device, as well as in the interfaces between associated powers. In fact, the range of variation of the voltage at the terminals of the supercapacitor is limited, in order to control the efficiencies. We introduce the discharge factor d, which is defined by the relation between the minimum voltage, U_m, below which the component cannot be discharged, and the maximum admissible voltage, U_M. This relation is expressed as a percentage:

$$d = \frac{U_m}{U_M} 100 \quad [9.4]$$

In fact, the total energy W_M, contained in the supercapacitor cannot be used, rather only a fraction known as the useful energy W_u:

$$W_u = W_M \left(1 - \left(\frac{d}{100} \right)^2 \right) \quad [9.5]$$

For example, for d=50% (the minimum voltage is half the maximum voltage, which defines the full charge), the useful energy that can be extracted from the supercapacitor is 75% of the total energy stored. In order to guarantee a good efficiency, we avoid choosing a d that is less than 50%.

From the definition of the useful energy that can be used for a supercapacitor, we can finally identify the number of devices, N, that is required to supply our energy needs W:

$$N = \frac{W}{W_m \left(1 - \left(\frac{d}{100} \right)^2 \right)} \quad [9.6]$$

For a given type of supercapacitor there is no unique solution to the number of devices. This number depends on the discharge factor, d. Therefore, we have a large degree of freedom in sizing a bank of supercapacitors.

9.3.2. *Power criteria – notion of efficiency*

As mentioned in the definition of a model for a supercapacitor, the model consists of a series resistance, which represents the internal losses during the charging and discharging phases. Taking into account these losses gives us a notion of the efficiency, which must be accounted for during the sizing of a bank of supercapacitors.

For example, in Figure 9.6 we give efficiency curves for a device with 2,600 F/2.5 V/0.7 mΩ, in charge and discharge, according to the constant current and constant power profile, according to a discharge factor of d=50%. We note that it is always possible to refer to such a profile, even if the charge/discharge currents and powers are not constant.

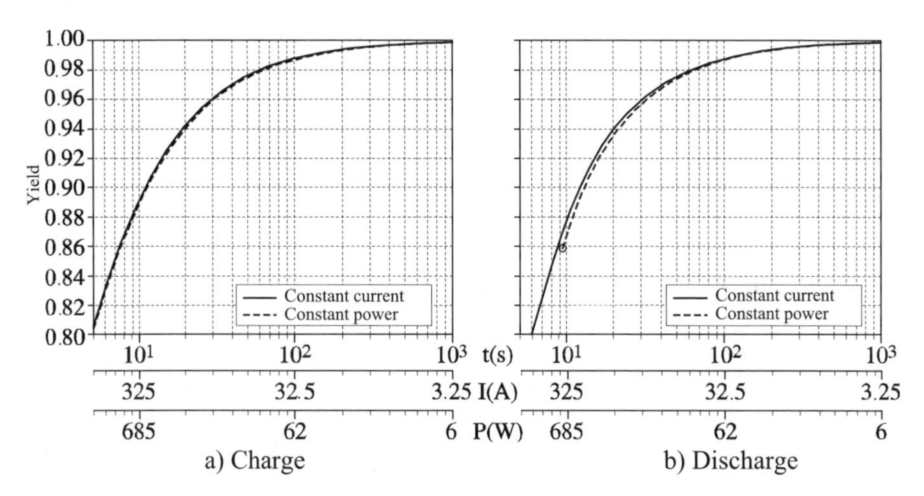

Figure 9.6. *Efficiency of a supercapacitor 2,600 F/2.5 V/0.7 mΩ, d=50%*

In the case of charging, as in the case of discharging, for the device considered, the currents and the powers must be limited to keep the efficiency at values that are greater than 90%, even though the series resistance is a low value (0.7 mΩ).

Thus, for charging, we must limit the current to 297 A, or the power to 604 W, to obtain an efficiency of 90%. The conditions are stricter for discharge, where the discharge current must be limited to 267 A, and the discharge power must be limited to 423 W.

If we take the last value into account, we can deduce the power density of the component, defined by the relation between the admissible discharging power of component to obtain 90% efficiency (423 W) and the weight of the component (0.525 kg). In the case of the device considered in our example, we obtain a power density of 806 W/kg, whereas the power density given by the manufacturer is 4,300 W/kg.

Given the gap between the measured performances and those claimed by the manufacturers, it is necessary to take into account the notion of efficiency in the sizing of a bank of supercapacitors, using the knowledge of the series resistance of the devices considered.

An example where the efficiency is taken into account is given in Figure 9.7 for the definition of a bank of supercapacitors dedicated to supplying 170 kJ under 30 kW from devices with 350 F/2.5 V/3.2 mΩ.

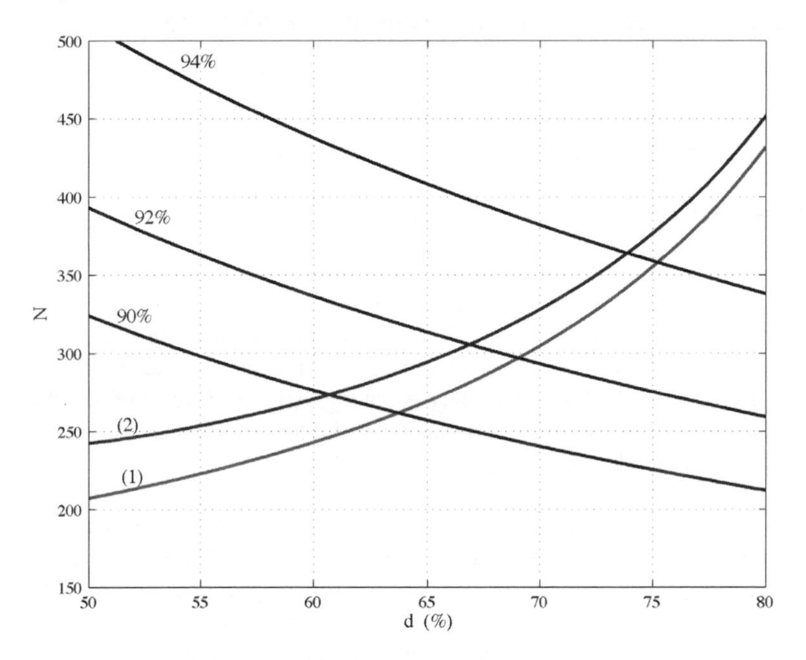

Figure 9.7. *Sizing of a bank of supercapacitors – taking into account the efficiency*

If we do not take into account the efficiency, the number of devices is identified using the energy needs according to equation [9.6]. There is no unique solution as for a given energy to be stored, the number of devices depends on the discharge factor *d*. For this factor varying from 50% to 80%, we obtain the locus of points (1) in Figure 9.7, which gives a family of possible choices.

We have equally shown in Figure 9.7 the locus of points of iso-efficiency under 30 kW (90%, 92%, and 94%). In this way, we see that the efficiency is rising with the number of devices and rising with the discharge factor.

Finally, the locus of points (2) in Figure 9.7 defines a family of possible choices to answer the energy criteria (170 kJ), taking into account the efficiency. This curve is closer to the ideal curve when the efficiency is close to 1 (for a high N, d is close to 100%). In all cases, taking into account the efficiency leads to an increase in the number of devices to be used, while limiting the range of variation of the voltage at the terminals of the devices. In the end, the determination of the number of devices to use is achieved by considering the intersection of the curve (2) (energy criteria with efficiency accounted for) with the curves of iso-efficiency (power criteria).

In the example examined here, the answer to specifications of 170 kJ/30 kW may be N=270 and d=60.8 % for an efficiency of 90%, or else N=365 and d =73.8 % for an efficiency of 74%.

9.4. Power interfaces

9.4.1. *Balancing voltages*

The example developed for the sizing of a bank of supercapacitors shows that the number of devices to be used can prove to be important. Moreover, the supercapacitor is a very low voltage source by nature. Having to link together a high number of devices is beneficial as putting components in series relieves the operating voltage of the bank. This enables, at a given power, reduction of the value of the charge/discharge current, and guarantees high performance efficiency for the linked static converter.

However, putting devices in series that do not strictly have the same capacitances (±20%) poses a problem in balancing the voltages between the different devices. At the end of the charging phase of the bank of supercapacitors, the voltages at the terminals of the least capacitance devices may exceed the maximum admissible voltages. If the charge is stopped as soon as at least one of the devices reaches its maximum voltage, the greater capacitance devices will see a lesser voltage.

Therefore, the energy stored in the latter devices is not at its maximum level, and we are, therefore, not able to exploit the entire energy capacity of the bank.

Therefore, the voltages of each of the devices put in series should be balanced so as to avoid exceeding the maximum admissible voltage and reducing the lifespan of the supercapacitors. Instead, where possible, we must exploit the levels of energy that can be attained.

Among the first devices proposed for balancing voltages, we give in Figure 9.8 an example solution, for an accumulator made up of eight Supercapacitors [BAR 00], [BAR 02].

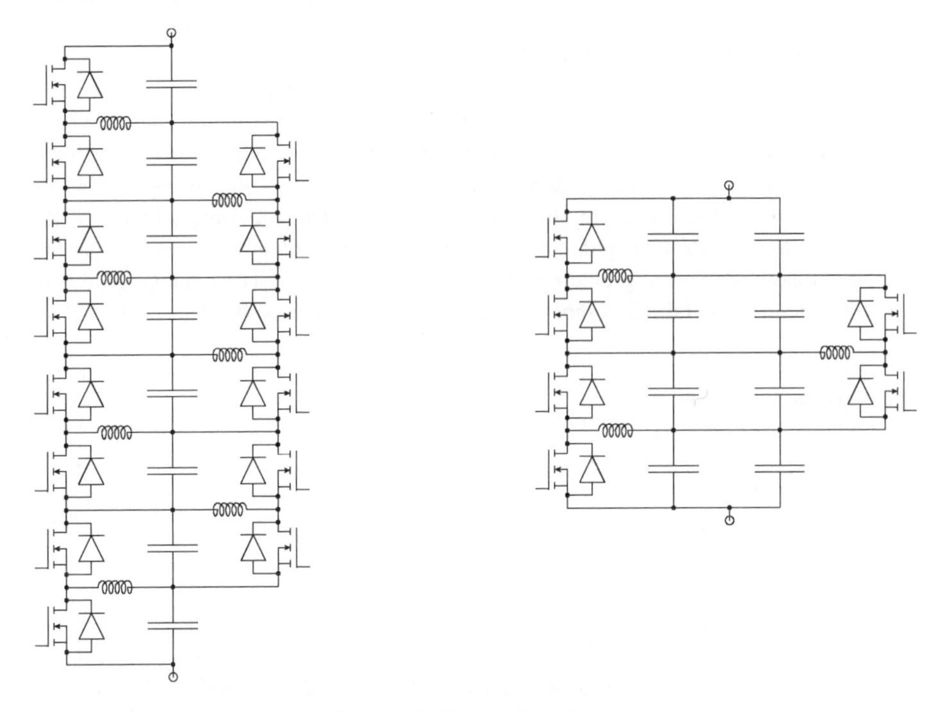

Figure 9.8. *Balancing voltages*

A reversible current *buck-boost* converter is linked to each pair of supercapacitors. As soon as a significant difference in voltage is detected between two successive supercapacitors, the associated converter is activated with a cyclic relation of one half. As soon as the voltages are balanced, the converter is deactivated. In order to guarantee the best possible efficiency, the converters are realized using MOSFETs with a low $R_{ds(on)}$, rapid recovery diodes, and function in discontinuous conduction mode.

Even if such a balancing system fits in terms of efficiency and of balancing dynamics, it is a complicated solution for which the number of components increases the higher the number of supercapacitor devices. However, the number of equalizers can be reduced if the eight devices are coupled in two parallel branches of four supercapacitors in series. An additional gain, in this case, comes from placing the devices in parallel, which reduces the difference in capacitances between the equivalent capacitors connected in series, which reduces the problems of voltage balancing and facilitates the equalization.

Nevertheless, the solutions that are used today are simpler. A resistance is associated with each device, which is only triggered if the voltage exceeds the admissible voltage. In that case we create an artificial leakage current with the aim of bringing the voltage of the device back to the maximum authorized value, with a slow dynamic (current of order of a hundred milli-amps). During the first charge/discharge cycles, some devices will, therefore, have voltages exceeding the maximum value. The voltages will be balanced after a few tens of cycles, at least when the bank is in its fully charged state. The balancing resistances are then no longer required. When they are again required, it is because the capacitance of the associated devices has changed. This is notably the case towards the end of the life of the component or during external destructive phenomena. The balancing devices can then identify the failing components that require replacement.

9.4.2. *Static converters*

Supercapacitors can be considered as sources of continuous voltage. However, the voltage at the terminals of a bank of supercapacitors is not constant, but depends on its state of charge. Moreover, the value of the charge/discharge current must be controlled in order to keep the efficiency at the specified value.

In fact, a bank of supercapacitors cannot be directly connected to its load. A static converter must be the interface between it and its application in order to upgrade the current/voltage values between the accumulator and its application. Moreover, the controller associated with the static converter should allow the control of the charge/discharge current.

A certain characteristic should be taken into account from the start: although the series resistance of a supercapacitor can be considered as a constant over a large range of frequencies, it has the capacity to fall drastically, in order to be considered as zero after around a hundred Hertz [BUL 02]. Although the average component of the charge/discharge current of the supercapacitor determines the energy storage, any harmonic component should be minimized to prevent the device from falling in the frequency range where it will only be resistive and therefore dissipative. The definition of the structure of a static converter should satisfy this property.

Having taken into account these characteristics, a possible structure type for a static converter is given in Figure 9.9, where the bank of supercapacitors is represented by an equivalent capacitance C.

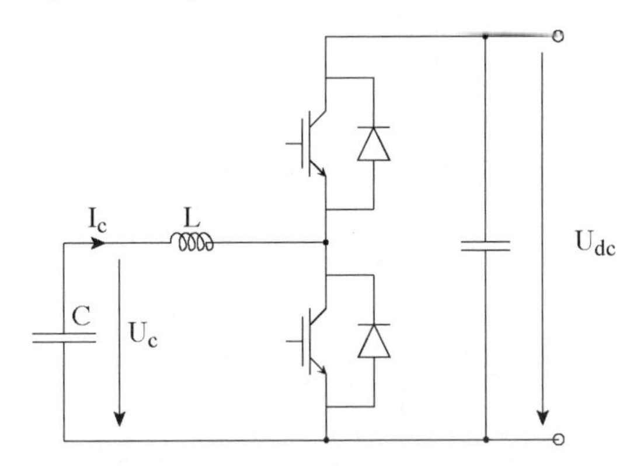

Figure 9.9. *Elevating voltage converter with reversible current*

This converter is a continuous-continuous converter with reversible current, which increases the voltage: the voltage generated, U_{dc}, is in all cases greater than the voltage at the terminals of the bank, U_c. This is important because it avoids the use of a high number of devices in series as supercapacitors are essentially low-voltage sources. In addition, the inductance, L, is sized to reduce undulations in the charge/discharge current, I_c, around its average value. This allows the high frequency requests on the supercapacitors, linked to converter cut-outs, to be reduced. This structure can be optimized in terms of efficiency, weight, and volume, thanks to interlaced canal technology, where each canal functions with discontinuous conduction.

It is the only structure that enables an increase in voltage as well as simultaneous direct control of the harmonic content of the current taken from the supercapacitors. Indeed, when an application demands a solution to specifications that the structure in Figure 9.9 cannot answer, it constitutes the first step of a cascade of converters as presented in Figure 9.10.

The structure presented in Figure 9.10a places in cascade a voltage-increasing converter with reversible current and a voltage-lowering converter with reversible current, via an intermediary stage at high voltage. Such a linking of converters is justified when the generated voltage U_{dc}, is susceptible to becoming smaller than the voltage at the terminals of the bank of supercapacitors, U_c.

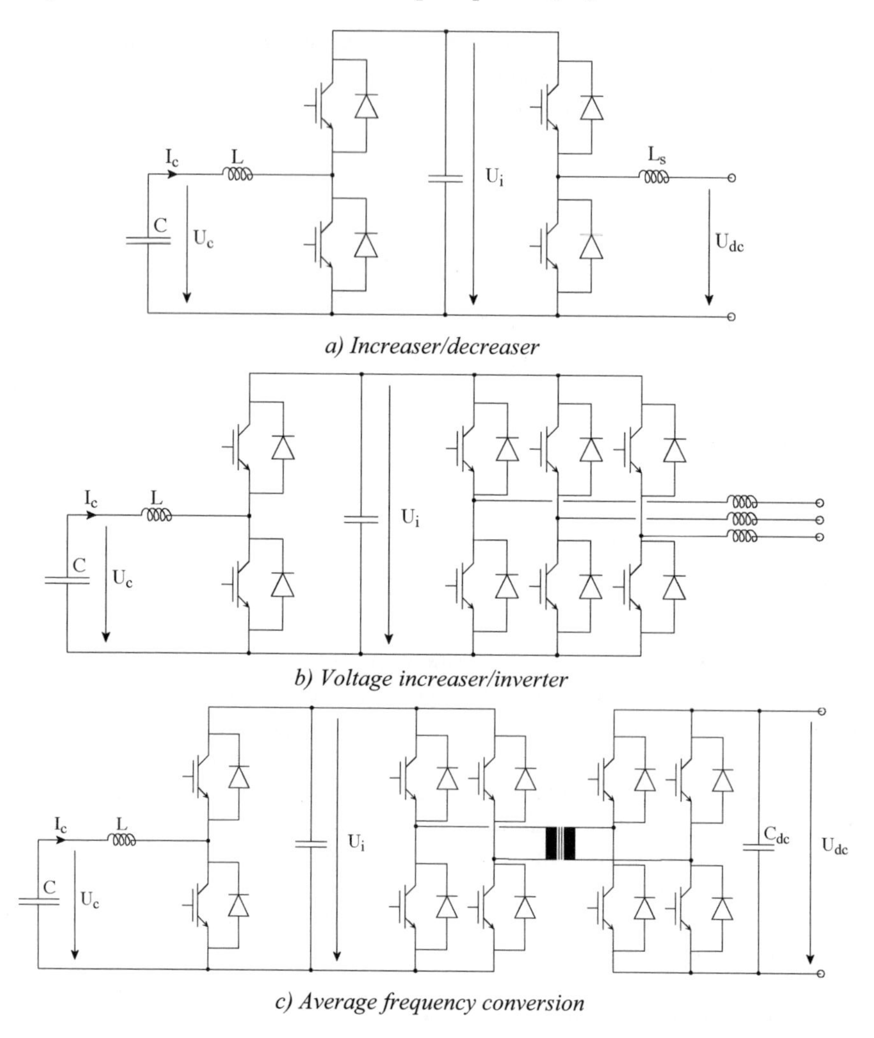

Figure 9.10. *Conversion chains*

The structure presented in Figure 9.10b is used when alternative values must be generated to supply an application (three-phase system in the presented case). The role of the continuous-continuous converter is to increase the voltage at the terminals of the bank of supercapacitors in order to supply a voltage inverter, for which the voltage U_i, should always be greater than the value of the voltages in the alternative phase. Here, the increased voltage stage is required because of the low-voltage nature of supercapacitors and in order to avoid the use of too many devices.

Finally, in the continuous-continuous conversion framework, the structure presented in Figure 9.1c can be used when the gap between the voltages U_c and U_{dc} is such that is cannot be assumed by a unique voltage-increasing converter (limits of the cyclic relation, regulation). Then we rely on an average frequency conversion. This principle consists of starting with the voltage U_i generated by a voltage increaser, and supplying the primary of a transformer with frequencies of the order of a kilohertz, using a stage inverter. The secondary of the transformer is supplied by a second voltage inverter, which functions as a rectifier permitting supply to the application. The use of a transformer in the kilohertz range enables part of the increase to be ensured while limiting the weight and volume of this element.

Finally, we should mention that all the structures presented in Figure 9.10 have a voltage-increasing converter as the entrance stage. This converter can be omitted for applications demanding voltage levels compatible with those directly proposed by the bank of supercapacitors (in its discharged state). This offers a gain in the efficiency, but a filter must be placed between the bank of supercapacitors and the voltage-lowering converter, the tension inverter or the average frequency converter, charged with limiting the harmonic content of the charge/discharge current [RUF 08].

9.5. Applications

9.5.1. *Generalities*

The application domains for supercapacitors are defined principally by three criteria: their energy density, their power density, and their lifespan (number of charge/discharge cycles).

The energy density of supercapacitors is low compared with that of batteries. The number of applications for which supercapacitors are enough to ensure a significant autonomy of energy is therefore reduced. However, the power density of supercapacitors, as well as their lifespan, enables their use in any application with reduced energy needs where the instantaneous power needs cannot be met by batteries.

We distinguish two types of applications. The first type consists of a unique principal source, charged with supplying power peaks of reduced duration. In the second application, which is more common, supercapacitors are used as complementary sources, in association with a principal energy source. The objective is to use supercapacitors to limit any power restrictions on the principal energy source. This could therefore be any hybrid system application.

9.5.2. *Supercapacitors used as principal source*

The number of applications for which supercapacitors are used as the principal source is low due to their energy density. Nevertheless, we can cite two key applications.

The first application is linked to the starting-up of internal combustion engines. For this type of application, energy needs are not important, so supercapacitors are adapted to answer instantaneous power supply needs of start-up. In addition, the high number of cycles characterizing supercapacitors is compatible with the lifespan of a thermal motor, which is an undeniable asset compared with batteries of accumulators, especially if we take into account the savings in terms of maintenance [SCH 00].

The second application is linked to supply of security functions, where it is necessary to supply a system during a short time, in case of a short general power-cut. We can list security supplies dedicated to computers, whose autonomy must at least allow the correct switching off of the computers. Generally, the autonomy sought is of the order of a minute to tens of minutes. The advantage of using supercapacitors instead of batteries resides in their lifespan, in the absence of maintenance operations, and in their rapid recharging when the principal power source is once again active.

The same type of applications include the use of supercapacitors associated with wind turbine blades. In the case where the power supply to the control systems is cut, the aim is to supply the controllers for the orientation of the blades in case of violent winds, as part of a safeguarding procedure for the wind turbine. In this case, the energy needs are compatible with the power density of supercapacitors. The high number of cycles and the absence of maintenance are equally predominant advantages. However, the penalizing factor is related to the leakage currents of supercapacitors, whose state of charge must, therefore, be monitored and maintained.

9.5.3. *Hybrid systems*

The power supply to hybrid systems is split in two, each of the sources supplying the application in a complementary manner. In the ideal case, the sources are associated in such a way so that each one is in its nominal usage range.

In hybrid systems, they are used principally because of their power density and their aptitude for surviving cycles. We use them as secondary sources in order to answer power needs or fluctuating power needs of an application. The principal energy source can be a supply network, batteries of accumulators [BAR 08], an internal combustion engine, a fuel cell [DIE 03], etc. The objective in these cases is to share the functionalities: the principal supply source satisfies the energy needs of the system, whereas the supercapacitors provide for the power needs.

The hybridization of a system with the help of supercapacitors as a secondary source covers a very wide spectrum of applications.

In the low-power range, supercapacitors are associated with batteries in devices, such as a camera or camcorder, in order to increase the lifespan of the batteries by reducing the restrictions that are applied to them.

As to applications with power in the order of tens of kilowatts, replacement of some of the braking resistances used in elevators by supercapacitors is being considered. The objective is to cover a wide range of the energy and power needs of the elevator using supercapacitors, and to reduce the impact on the supply network of the up/down cycles of the elevator, which leads to discontinuous power profiles [RUF 02].

Automobile and traction sectors are target sectors for supercapacitor technology. With regards to automobiles, the functions naturally answered by supercapacitors are of the *stop and go* type, or much more advanced functions that reduce the power constraints on the internal combustion engine.

With regards to traction, several possibilities for using supercapacitors have been developed in parallel. A first application concerns the supply to tramways or trolleybuses, using catenaries. It is possible to use supercapacitor technology to reduce the impact on the catenary of the power profiles demanded by the vehicles, while allowing braking by recovery to be exploited as much as possible. Several solutions are tested today: connection of a bank of supercapacitors via its power interface on the catenary (control of the catenary voltage, storage of the energy supplied by the braking of vehicles) [RUF 04b], [SIT 00], or integration of supercapacitors directly on the vehicle (local reduction of the constraints on the catenary, braking by recovery) [DES 07], [STE 04]. A second application is the hybridization of diesel-electric regional trains, an example of which is presented in Figure 9.11.

A study was undertaken for the vehicle presented in Figure 9.11a: a diesel electric train developed by the Stadler Rail AG [DES 04]. The train studied was configured for the Merano-Malles line (in northern Italy), characterized by steep gradients: altitudes of the line being considered vary between 1,000 m and 1,700 m. As presented in Figure 9.11b, this train is initially supplied by two diesel generators, each capable of delivering 380 kW (760 kW in total). It should be noted that in such a configuration, these diesel generators must supply all the fluctuating power needs of the vehicle. Therefore, the engines are governed by frequent changes in the regime, which makes controlling their efficiency and their emissions difficult. In addition, braking with energy recuperation is not possible: the energy supplied by the diesel generators can never be capitalized.

a) Diesel-electric train

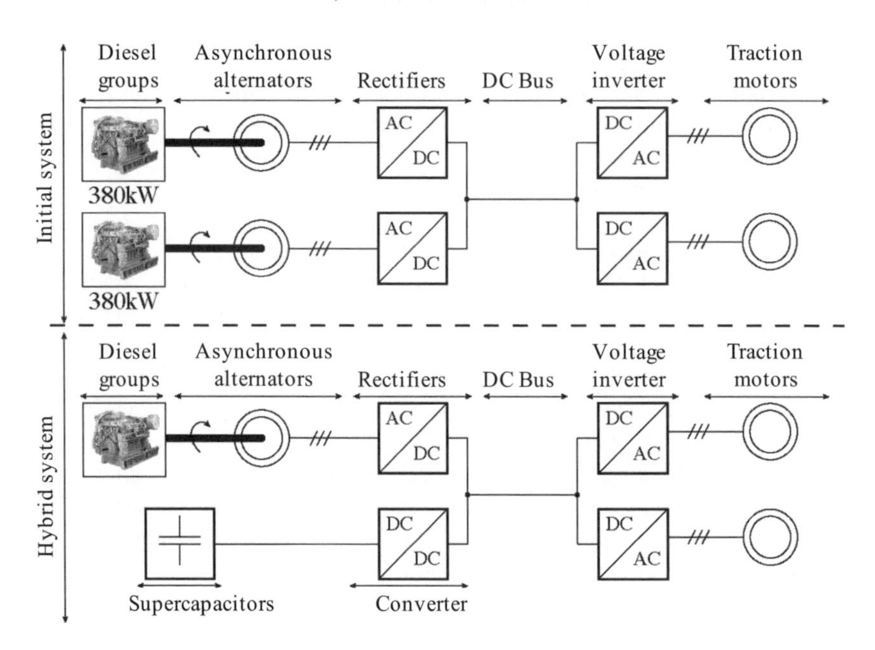

b) From the initial system towards its hybridization

Figure 9.11. *Hybrid vehicle – sizing of the diesel generator*

The proposed study sizes a bank of supercapacitors and integrates them on that train, as presented in Figure 9.11b, which should allow some or all of the fluctuating power needs of the vehicle to be answered. Indeed, the objective is to control the diesel generator in a stop/go mode. During the go phases, the diesel generator is maintained in an operational mode, in a regime offering the best efficiency and reduced emissions. Moreover, storage using supercapacitors ensures that there will be energy recuperation functions during braking in order to increase the efficiency of the system.

The evaluation of how much energy to store in supercapacitors in order to satisfy these criteria is given in Figure 9.12, and shows its direct influence on the diesel power that must be installed.

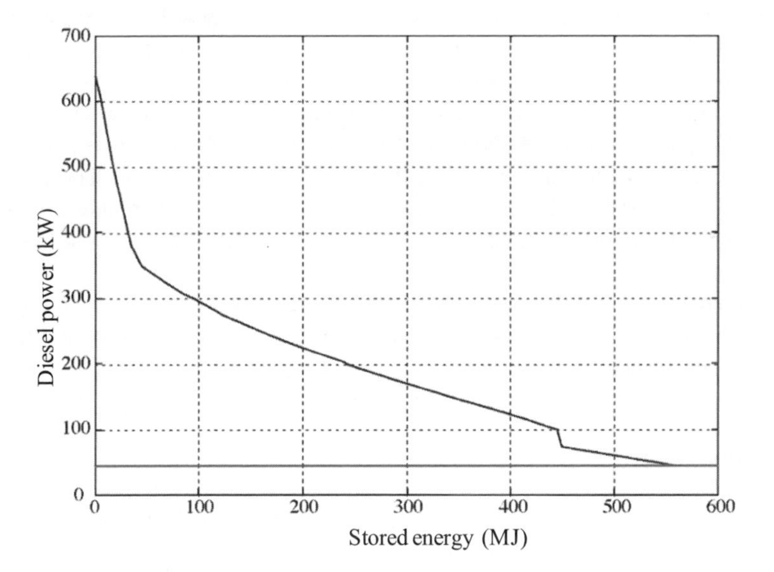

Figure 9.12. *Impact of storage on the diesel power installed*

The reference point is the absence of storage (zero energy stored), which defines the power initially predicted (640 kW). An extreme point shows that if it is possible to define a bank of supercapacitors that is compatible with the weights and volumes offered by the vehicle, and capable of storing 550 MJ, then only 45 kW of diesel power needs to be installed. Unfortunately, this choice is not possible because of the dimensions that the storage device would have to have.

An interesting choice consists of defining a bank of supercapacitors that can store an energy of 45 MJ. Therefore, the initial power supplied by the diesel generator would reduce by half, so that one of the diesel engines could be retired and substituted by a bank of supercapacitors made up of 4,900 devices with 3,000 F/2.7 V/0.29 mΩ, which has a weight of 2.7 metric tons and a volume of 2.3 m^3, as shown in Figure 9.11b.

The gain, which is deduced from calculations and simulations, is a reduction in fuel consumption of the order of 44%, and a return on investment in 10 years, linked to the additional cost of the supercapacitor technology. We should note that the calculations were carried out at a time when a barrel of petrol was not yet expected to cost more than US$100.

As final applications, we will cite those that are related to the principle of boost charging [RUF 04a]. This is about giving a vehicle a reduced range, using a bank of supercapacitors. Moving along a fixed route, the onboard supercapacitors are recharged when the vehicle stops at one of the stations distributed along its journey. As there must be a rapid transfer of energy from the station to the onboard accumulator, the powers involved can prove to be significant (from 100 to 1,000 kW), during a reduced duration (tens of seconds) – the energy is supplied by a bank of supercapacitors integrated within the recharging station. These are then recharged between the vehicles' arrivals. Benefiting from a few minutes, the powers taken from the network are then no more than of the order of dozens of kilowatts.

Such an application illustrates the major role that supercapacitors can play in the domain of energy storage: used as buffer reservoirs, they must limit the power demands on a principal energy source.

9.6. Bibliography

[BAR 00] BARRADE P., PITTET S., RUFER A., "Series connection of supercapacitors, with an active device for equalizing the voltages", *PCIM 2000: International Conference on Power Electronics, Intelligent Motion and Power Quality*, June 6-8, Nurnberg, Germany, 2000.

[BAR 02] BARRADE P., "Series connection of supercapacitors: comparative study of solutions for the active equalization of the voltages", *Electrimacs 2002, 7th International Conference on Modeling and Simulation of Electric Machines, Converters and Systems*, August 18-21, Montreal, Canada, 2002.

[BAR 03a] BARRADE P., "Energy storage and applications with supercapacitors", *ANAE: Associazione Nazionale Azionamenti Elettrici, 14o Seminario Interattivo, Azionamenti elettrici: Evoluzione Tecnologica e Problematiche Emergenti*, March 23-26, Brixen, Italy, 2003.

[BAR 03b] BARRADE P., RUFER A., "Current capability and power density of supercapacitors: considerations on energy efficiency", *EPE 2003: European Conference on Power Electronics and Applications*, September 2-4, Toulouse, France, 2003.

[BAR 08] BARRADE P., DESTRAZ B., HAUSER S., RUFER A., "Application de supercondensateurs dans le transport individuel – étude expérimentale d'un scooter électrique avec assistance en puissance", *Bulletin de l'Association pour l'électrotechnique, les technologies de l'énergie et de l'information et de l'Association des entreprises électriques suisses (SEV/AES)*, pp. 37-41, 2008.

[BUL 02] BULLER S., Impedance-based simulation models for energy storage devices in advanced automotive power systems, PhD thesis, ISEA, 2002.

[DES 04] DESTRAZ B., BARRADE P., RUFER A., "Power assistance for diesel - electric locomotives with supercapacitive energy storage", *IEEE-PESC 04: Power Electronics Specialist Conference*, June 20-25, Aachen, Germany, 2004.

[DES 07] DESTRAZ B., BARRADE P., RUFER A., KLOHR M., "Study and Simulation of the Energy Balance of an Urban Transportation Network", *EPE 2007: 12th European Conference on Power Electronics and Applications*, September 2-5, Aalborg, Denmark, 2007.

[DIE 03] DIETRICH P. *et al.*, *Hy. Power* – "A technology platform combining a fuel cell system and a supercapacitor", in *Handbook of Fuel cells – Fundamentals, Technology and Applications*, vol. 4, part 11, pp. 1184-1198, John Wiley & Sons, Chichester, 2003.

[RUF 02] RUFER A., BARRADE P., "A supercapacitor-based energy-storage system for elevators with soft commutated interface", *IEEE Transactions on Industry Applications*, vol. 38, pp. 1151-1159, 2002.

[RUF 04a] RUFER A., BARRADE P., HOTELLIER D., HAUSER S., "Sequential supply for electrical transportation vehicles: properties of the fast energy transfer between supercapacitive tanks", *Journal of Circuits, Systems and Computers*, vol. 13, pp. 941-955, 2004.

[RUF 04b] RUFER A., HOTELLIER D., BARRADE P., "A supercapacitor-based energy-storage substation for voltage-compensation in weak transportation networks", *IEEE Transactions on Power Delivery*, vol. 19, pp. 629-636, 2004.

[RUF 08] RUFER A., BARRADE P., CORREVON M., WEBER J.-F., "Multiphysic modeling of a hybrid propulsion system for a racecar application", *Iamf EET-2008: European Ele-Drive Conference, International Advanced Mobility Forum*, March 11-13, Geneva, Switzerland, 2008.

[SCH 00] SCHNEUWLY A., GALLAY R., "Properties and applications of supercapacitors from the state-of-the-art to future trends", *Power Conversion and Intelligent Motion Conference*, PCIM, Nurnberg, Germany, 2000.

[SIT 00] SITRAS SES Energiespeichersystem für 750V DC Bahnanlagen, Siemens Transportation Systems Public. Nr A19100-V300-B276 and B275, field of power electronics, and energy management for UPS applications. Patent application.

[STE 04] STEINER M., SCHOLTEN J., "Energy storage on board of DC fed railway vehicles", *35th IEEE PESC Conference*, Aachen, Germany, June 21-242004.

[ZUB 00] ZUBIETA L., BONERT R., DAWSON F., "Considerations in the design of energy storage systems using double-layer capacitors", *IPEC Tokyo*, Japan, p. 1551, 2000.

List of Authors

Philippe BARRADE
EPFL
Lausanne
Switzerland

Régine BELHOMME
EDF R&D
Clamart
France

Yves BRUNET
INP
Grenoble
France

Denis CANDUSSO
INRETS-LTN
Belfort
France

Orphée CUGAT
G2Elab
Grenoble
France

Jérôme DELAMARE
G2Elab
Grenoble
France

Gauthier DELILLE
EDF R&D
Clamart
France

Jérôme DUVAL
EDF R&D
Clamart
France

Daniel FRUCHART
Institut Neel
Grenoble
France

Florence FUSALBA
CEA-INES
Le Bourget-du-Lac
France

Daniel HISSEL
FEMTO
Franche-Comté University
Belfort
France

Jean-Marie KAUFFMANN
IGE
Franche-Comté University
Belfort
France

Gilles MALARANGE
EDF R&D
Clamart
France

Julien MARTIN
EDF R&D
Clamart
France

Sébastien MARTINET
CEA Liten
Grenoble
France

Florence MATTERA
CEA-INES
Le Bourget-du-Lac
France

Andrei NEKRASSOV
EDF R&D
Clamart
France

Marie-Cécile PERA
FEMTO
Franche-Comté University
Belfort
France

Eric VIEIL
LEPMI
Grenoble
France

Index

B-C

D

E-F

H-I

M-O

P

R

S

T-U

V-W